I0063332

Rietveld Refinement in the Characterization of Crystalline Materials

Rietveld Refinement in the Characterization of Crystalline Materials

Special Issue Editor

Igor Djerdj

MDPI • Basel • Beijing • Wuhan • Barcelona • Belgrade

MDPI

Special Issue Editor
Igor Djerdj
Josip Juraj Strossmayer University of Osijek
Department of Chemistry
Croatia

Editorial Office
MDPI
St. Alban-Anlage 66
4052 Basel, Switzerland

This is a reprint of articles from the Special Issue published online in the open access journal *Crystals* (ISSN 2073-4352) in 2018 (available at: https://www.mdpi.com/journal/crystals/special_issues/ rietveld_refinement)

For citation purposes, cite each article independently as indicated on the article page online and as indicated below:

LastName, A.A.; LastName, B.B.; LastName, C.C. Article Title. *Journal Name* **Year**, *Article Number*, Page Range.

ISBN 978-3-03897-527-4 (Pbk)
ISBN 978-3-03897-528-1 (PDF)

© 2019 by the authors. Articles in this book are Open Access and distributed under the Creative Commons Attribution (CC BY) license, which allows users to download, copy and build upon published articles, as long as the author and publisher are properly credited, which ensures maximum dissemination and a wider impact of our publications.

The book as a whole is distributed by MDPI under the terms and conditions of the Creative Commons license CC BY-NC-ND.

Contents

About the Special Issue Editor

Igor Djerdj (born on 23 August 1972 in Osijek, Croatia) studied physics at the University of Zagreb, Croatia where he also received his Ph.D. degree in 2003. He completed his first postdoctoral studies at the Max Planck Institute of Colloids and Interfaces in Potsdam, Germany. He then moved to the Swiss Federal Institute of Technology (ETHZ), Switzerland, where he completed his second post-doc. Since 2009 he has been a Scientific Associate at the Ruer Bošković Institute, Croatia, and in 2016 he moved to the J. J. Strossmayer University of Osijek where he was appointed associate professor. His research interests include the structural characterization of a variety of materials with targeted applications, theoretical modelling of the electronic structure, liquid-phase synthesis of novel materials, particularly inorganic–organic hybrids, diluted magnetic semiconductors, magnetic nanoparticles, and gas-sensing materials. His research results and achievements have been published in scientific journals, with a total of 86 peer-reviewed publications that have received more than 3000 citations and h = 31.

Preface to "Rietveld Refinement in the Characterization of Crystalline Materials"

This Special Issue serves as a crystallographic forum covering various aspects of the material science that have in common the use of the powerful Rietveld method in the analysis of powder XRD patterns of the investigated compounds. It consists of seven papers describing diverse research topics: The article by Altomare et al. presents the most recent solution strategies in their software EXPO, both in reciprocal and direct space, aiming at obtaining models suitable to be refined by the Rietveld method. Skoko et al. report in their contribution an interesting phenomenon, the thermosalient effect, observed on methscolopamine bromide. Yakimov et al. propose a self-configuring genetic algorithm that provides a fully automatic analysis of the electrolyte composition by the Rietveld method including successful testing examples. The Rietveld analysis with the direct space method for the structural determination of binary tetrahydrofuran (THF) + O2 and 3-hydroxytetrahydrofuran (3-OH THF) + O2 clathrate hydrates is another structural challenge considered in this Special Issue and was co-authored by Ahn et al. Zibrov and Filonenko utilize the Rietveld method in the structural analysis of boron-doped diamonds. Chatelier et al. studied precipitates in thin-walled and thick-walled microalloyed X70 pipe steel using Rietveld refinement and quantitative X-ray diffraction. Finally, Zhao et al. researched error analysis and correction for quantitative phase analysis based on the Rietveld-internal standard method. The results presented in the articles collected in this Special Issue clearly demonstrate that Rietveld refinement occupies a high place in the advanced structural characterization of crystalline materials.

Igor Djerdj
Special Issue Editor

crystals

MDPI

Article

Error Analysis and Correction for Quantitative Phase Analysis Based on Rietveld-Internal Standard Method: Whether the Minor Phases Can Be Ignored?

Piqi Zhao [1], Lingchao Lu [1], Xianping Liu [2], Angeles G. De la Torre [3] and Xin Cheng [1],*

[1] Shandong Provincial Key Laboratory of Preparation and Measurement of Building Materials, University of Jinan, Jinan 250022, China; mse_zhaopq@ujn.edu.cn (P.Z.); mse_lulc@ujn.edu.cn (L.L.)
[2] School of Materials Science and Engineering, Tongji University, Shanghai 201804, China; lxp@tongji.edu.cn
[3] Departamento de Química Inorgánica, Cristalografía y Mineralogía, Universidad de Málaga, 29071 Málaga, Spain; mgd@uma.es
*   Correspondence: chengxin@ujn.edu.cn; Tel.: +86-531-8276-7217

Received: 26 December 2017; Accepted: 19 February 2018; Published: 27 February 2018

Abstract: The Rietveld-internal standard method for Bragg-Brentano reflection geometry ($\theta/2\theta$) X-ray diffraction (XRD) patterns is implemented to determine the amorphous phase content. The effect of some minor phases on quantitative accuracy is assessed. The numerical simulation analysis of errors and the related corrections are discussed. The results reveal that high purity of crystalline phases in the standard must be strictly ensured. The minor amorphous or non-quantified crystalline phases exert significant effect on the quantitative accuracy, even with less than 2 wt% if ignored. The error levels are evaluated by numerical simulation analysis and the corresponding error-accepted zone is suggested. To eliminate such error, a corrected equation is proposed. When the adding standard happens to be present in sample, it should be also carefully dealt with even in low amounts. Based on that ignorance, the absolute and relative error equations (Δ_{AE}, Δ_{RE}) are derived, as proposed. The conditions for high quantitative accuracy of original equation is strictly satisfied with a lower amount of standard phase present in sample, less than 2 wt%, and a higher dosage of internal standard, larger than 20 wt%. The corrected equation to eliminate such quantitative error is suggested.

Keywords: Rietveld; quantitative analysis; corrected equation; amorphous

1. Introduction

Quantitative phase analysis based on X-ray diffraction (XRD) can be traced back to as early as 1919 [1]. Hull firstly proposed that this technique had the potential to perform accurately quantitative analysis. Subsequently, Alexander developed practical XRD methods and derived the related theoretical basis [2]. With the improvements of XRD analysis, different quantitative methods have been presented in succession, such as reference intensity ratio (RIR) method [3], external standard method [4], matrix-flushing method [5], non-standard method [6], doping method [7], and Rietveld method [8]. The Rietveld method, as proposed by Hugo M. Rietveld, is widely accepted due to its whole-pattern fitting approach instead of single-peak analysis. The main advantage is that it can effectively minimize or eliminate the inaccuracies arising from preferred orientation, particle statistics, microabsorption, peaks overlapping, and detection of amorphous phase and trace phases [9–11]. Over the last two decades, it has become widely accepted by scientific community, gradually being a standard practice, as it is possible to solve those problems associated with crystalline materials. However, there are two preconditions: (1) the quantified phase is the crystalline phase, and (2) the crystal structure is known [12]. Therefore, if amorphous material is present, quantitative results could not be obtained directly. In such case, it is normal to use a reference material for the recalculation of the phase contents [13]. This reference material is either mixed in the sample as internal standard [10] or measured separately under identical conditions as external standard [14]. The internal standard method derives the amorphous content from the comparison between the actual dosage and Rietveld result of the internal standard, Equation (1) [13]:

$$W_{Amor} = \frac{1 - W_{St}/R_{St}}{100 - W_{St}} \times 10^4 \tag{1}$$

where W_{St} stands for the actual dosage of the internal standard, and R_{St} stands for the Rietveld results of the internal standard.

Meanwhile, the external standard method focuses on determining the diffractometer constant with an appropriate standard, which is used to determine the weight fraction of each crystalline phase. From the difference between 100 wt% and the sum of the crystalline phase contents, an overall weight percentage of amorphous can be subsequently derived.

$$G = K_e = S_{St} \cdot \rho_{St} V_{St}^2 \mu_{St} / W_{St} \tag{2}$$

$$W_{Amor} = 100 - \sum S_\alpha \cdot \rho_\alpha V_\alpha^2 \mu_s / G \tag{3}$$

Where S_{St} and S_α stands for the scale factors of the external standard and each phase in the mixture, respectively, ρ is the density, V is the unit cell volume, and μ_{St} and μ_s are the mass absorption coefficient of the external standard and the sample, respectively.

Besides these two methods, the amorphous content could also be quantified by the 'PONKCS' (Partial Or No Known Crystal Structure) method that relies upon treating a set of peaks of amorphous phase as a single entity [15]. Amorphous phase is characterized by measured rather than a calculated structure factor.

The Rietveld-internal standard method, as the most widely used technique, is a relatively easy and direct strategy, no other measurements or calibrations are needed. If the analyses are carefully performed and the amount of amorphous content to be determined over 15 wt%, the accuracy is satisfactory, close to 1% [10,13,16–18]. On the contrary, the external and 'PONKCS' methods are more complicated and some other preconditions need to be met in the meantime. For instance, the external standard method strictly requires the identical conditions of XRD data acquisition and different mass attenuation coefficient correction between sample and standard. Though 'PONKCS' strategy may solve the problem of undistinguished ability between different amorphous phases, it depends on accurate identification and the calibration of amorphous. A problem may occur because of the complexity of this analyzing process, especially when the amorphous phases are less evident or difficult for characterization [19].

Beside the advantage of the internal method mentioned above, the determination of amorphous contents is also a very challenging operation indeed. First of all, the effects of internal standard should be well considered. A significant error may occur when the selected standard is not homogeneously mixed into the sample or when an obvious absorption contrast exists between sample and standard [20]. Furthermore, use of an appropriate amount of internal standard is also a key point to guarantee the accuracy. Discussions about that influence on accuracy of amorphous phase quantitation have been reported [21,22]. The results illustrated that the quantitative accuracy of amorphous phase follows a nonlinear function by Rietveld-internal standard method, which in turn leads to a serious error in determining the minor amount of amorphous content. Most of these issues can be mitigated based upon the above results by adequate sample preparation and correct data acquisition [23–26]. However, there are still some factors that need further attention. Here, we study the error analysis that was introduced by some minor phases that are related to the internal standard. They include the minor impurity phases of the internal standard and the sample containing the same crystalline phase as the internal standard. There is often an overlook about contribution of these minor phases. The basic goal is to understand such effect on the quantitative accuracy. In order to do so, a three-dimensional numerical simulation database, where the information includes absolute and relative errors is conducted, and relative corrected equations have been proposed. This study is a step forward to better understand quantitative phase analysis based on Rietveld-internal standard method.

2. Materials and Methods

2.1. Raw Materials

The Powders of SiO_2 [ABCR GmbH. Co. KG (Karlsruhe, Germany)] and ZnO [Sigma-Aldrich, Co. LLC (St. Louis, MO, USA)], purity of 99.9%, are chosen as the internal standard in this work. They were sieved firstly through 74 μm prior to be used. The glass powder was adopted as amorphous component, ground with particle sizes less than 20 μm (80% in number statistics), and surface area 500 m^2/kg determined by specific surface area measuring instrument. The glass has a chemical composition (wt% of oxides) determined by XRF, of SiO_2 (72.5), Al_2O_3 (1.8), CaO (5.8), Na_2O (13.3), K_2O (1.6), MgO (3.8), and ZnO (0.8).

2.2. Sample Preparation

The powders of SiO_2, ZnO and glass were weighed, with a designed mass ratio of 45%:45%:10%, as 4.507 g, 4.473 g, and 1.033 g, respectively. The above powders were subsequently mixed and homogenized by hand for 30 min in an agate mortar. Finally, they were uniformly filled into the holder and slightly leveled for XRD measurement. That mixtures were prepared and then underwent XRD tests by triplicate. XRD patterns didn't show significant differences.

2.3. Data Collection and Processing

Chemical composition of glass was determined by the X-Ray Fluorescence (SRS3400, Bruker AXS Corporation, Karlsruhe, Germany). Particle statistic of powders was measured by laser particle size analyzer (LS 230 from Beckman Coulter, Brea, CA, USA). The X-ray diffraction patterns of mixtures were measured in Bragg-Brentano reflection geometry ($\theta/2\theta$) on an X'Pert MPD PRO diffractometer (PANalytical International Corporation, Almelo, Netherland) and GSAS-EXPGUI software (Los Alamos National Laboratory, Los Alamos, NM, USA) [27]. The detailed instrument settings for XRD are summarized in Table 1.

Table 1. The Instrument Settings for X-ray Diffraction (XRD).

Scanning Type Detector	Continuous Scanning X'Celerator Detector
Geometry	Reflection/flat sample
X-ray radiation/tube working conditions	$CuK\alpha_1$, 45 kV/40 mA
Primary Monochromator	Ge (111)
Anti-scatter slit/°	1/2
Soller slit (rad)	0.04
Divergence slit/°	1/2
Angular range, 2θ/°	5–70
Step width/°	0.0167
Measure time/h	2
Sample spinning speed (r.p.m)	15

3. Results and Discussion

3.1. Quantitative Error Induced by Minor Impurity Phase of Internal Standard

The XRD pattern of that mixture was analyzed by Rietveld whole-pattern fitting based upon GSAS-EXPGUI software. The starting crystalline structure models of SiO_2 [28] and ZnO [29] were imported from literature. The instrument function file was chosen based on $CuK\alpha_1$ as the incident

X-ray and Germanium as the monochromator (monochromatic model with wavelength of 1.54056 and polarization fraction value of 0.8). For the refinement of peak shape parameters, pseudo-Voigt function [30] with asymmetry correction [31] was chosen and the peak width and asymmetry factor, such as LY, GW, H/L, and S/L were initially set to 12 (0.01°), 5 (0.01°), 0.02, and 0.02, respectively. The refined parameters included unit-cell parameters, zero-shift correction, background parameters, phase fractions, and peak shape parameters (LY and GW). A linear interpolation function was chosen to fit the background with polynomial term gradually increasing to 36. Peak shapes were fitted by refining the Gaussian contribution and Lorentzian contribution separately when appropriated. During the process of Rietveld refinement, the refined parameters had regular convergence and least-square R factors, assessing the fitness of pattern, decreased gradually. Figure 1 shows the Rietveld refinement pattern of the artificial mixture. The stable refinements and satisfactory fits, as indicated by the smoothness of the Yobs-Ycalc curve illustrated the Rietveld refinement was reliable. Moreover, the analysis was performed by triplicate in order to assess the precision. The results are close but not identical with relative errors lower than 1%. The quantitative results are listed in Table 2, which also includes the quantitative phase analysis corrected by taking into account the microabsorption effect [32,33]. This is important in this mixture as linear absorption coefficient for SiO_2 is 92 cm^{-1}, while that for ZnO is 290 cm^{-1}. Both standards have very similar particle size (~4 μm); consequently, ZnO will always be underestimated.

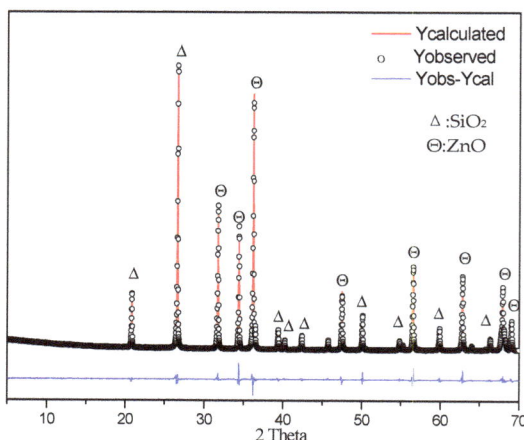

Figure 1. Rietveld XRD pattern of an artificial mixture (The circles correspond to observed data, the thin line is the calculated patterns by the Rietveld method. The Yobs-Ycal stands for the difference pattern plotted as blue line at the bottom). Main peaks due to each phase have been labeled Δ:SiO_2; ϴ:ZnO.

Table 2. Rietveld Quantitative Phase Analysis of the Mixture.

Phases	Weighed/wt%	SiO_2_ZnO_Glass	
		Riet/wt%, Uncorrected	Riet/wt%, Microabsorption Corrected
SiO_2	45.01	51.1	48.4
ZnO	44.67	48.9	51.6
Glass	10.32	——	——

The middle column contains direct Rietveld results, assuming that all of the phases in the sample are crystalline phases; the right-most column contains Rietveld results corrected for microabsorption.

To obtain the quantitative results of the amorphous phase (Glass), SiO_2 and ZnO were considered as internal standards, respectively. If microabsorption effect is not considered, the errors in amorphous determination are inevitably higher. However, even when considering that factor, the quantitative results are still not well reproducible with a significant relative deviation in the three different analysis, independently of the internal standard used (Table 3). However, when compared with different recorded XRD pattern analysis based on the same internal standard, the quantitative results show good consistency with absolute deviation of phases less than 1%.

Table 3. Rietveld quantitative results, including amorphous component derived by internal method for the artificial mixture, using the corrected values of Table 2.

Phases	Weighed/wt%	SiO_2_ZnO_Glass	
		[1]Sta(SiO_2)/wt%	[2]Sta(ZnO)/wt%
SiO_2	45.01	Fixed	41.90
ZnO	44.67	47.99	Fixed
Glass	10.32	7.00	13.43

'1' and '2' represents quantitative results included amorphous phase (glass), taking SiO_2 and ZnO as internal standard, respectively.

To further verify this conjecture, Rietveld quantitative phase analysis of that mixture was performed by external standard method (G-factor method). For calculating the G-factor, the polished polycrystalline quartz rock was firstly considered as the standard. It has the advantage of avoiding the error induced by powder standard during the sample preparation, such as fluctuation of surface roughness, packing density, and so on. XRD data collection of the quartz rock standard was as close in time and identical in diffractometer configuration as possible to the artificial mixture sample, which could make sense of the formula of 'G_SiO_2 = G_sample'. The mass absorption coefficient of artificial mixture sample under the condition of CuKα_1 radiation was determined as 42.34 cm^2/g by Highscore Plus software (PANalytical, Almelo, Netherland). After correlative parameters obtained from Rietveld refinement and crystallinity of quartz rock obtained from former analysis, G value was calculated as 5.51×10^{-20} (Table 4). Based on this G value, quantitative results of crystalline and amorphous phases were given in Table 5. The average quantitative result of 'Glass' phase among three calculations is 13.08%, which is about 3% larger than the original weighed fraction. The extra part is contributed by amorphous or non-quantified crystalline phases (ACn) in SiO_2 and ZnO.

Table 4. G Value Calculation Based on Rietveld Refinement of Quartz Rock.

Cquartz	Density [g/cm^3]	Refined Unit Cell Volume [cm^3]	Total Mass Absorption Coefficient [cm^2/g]	G Value
87.9	2.646	1.13×10^{-22}	34.84	5.51×10^{-20}

Table 5. Comparison of the Weight Fractions and Rietveld Quantitative Result (G-factor Method) of Artificial Mixture.

Phases	Weighed/%	SiO_2_ZnO_Glass
SiO_2	45.01	44.16
ZnO	44.67	42.75
Glass	10.32	13.08

To eliminate such quantitative errors induced by ACn in the internal standard, the original Equation (1) needs to be improved. Using Rietveld refinement, the improved equation for weight fraction of amorphous phase in original sample (W_{Amor}) can be derived as Equation (4), and the intermediate derivation process was shown in the supplementary materials.

$$W_{Amor} = \left[1 - \frac{(\frac{100}{R_{St}} - 1) \times W_{St} \times \alpha}{100 - W_{St}} \right] \times 100 \qquad (4)$$

where 'α' is defined as the crystallinity in that internal standard. The equations of absolute error are successively derived for theoretical calculation of the error level between the original (Equation (1)) and improved equation (Equation (4)).

$$W_{Amor(impr)} - W_{Amor(\text{origi})} = \frac{\left(\frac{1}{R_{St}} - \frac{1}{100}\right)}{\left(\frac{1}{W_{St}} - \frac{1}{100}\right)} \times (1 - \alpha) \tag{5}$$

To simplify Equation (5), R_{St} is firstly derived and expressed by crystallinity of internal standard (α) and the original sample (β) as Equation (6). The absolute error (Δ_{AE}) can be finally converted as Equation (7). Moreover, the relative error (Δ_{RE}) can be proposed as Equation (8).

$$R_{st} = \frac{W_{St}\alpha}{W_{St}\alpha + (100 - W_{St})\beta} \times 100 \tag{6}$$

$$\Delta_{AE} = W_{Amor(impr)} - W_{Amor(\text{origi})} = \frac{1 - \alpha}{\alpha} \times \beta \tag{7}$$

$$\Delta_{RE} = \frac{W_{Amor(impr)} - W_{Amor(\text{origi})}}{W_{Amor(impr)}} = \frac{1 - \alpha}{\alpha} \times \frac{\beta}{1 - \beta} \tag{8}$$

To evaluate the error level, the numerical simulation analysis of absolute and relative errors was performed. The results, corresponding to crystallinity of the internal standard (α) and the weight fractions of crystalline phases in original sample (β), were displayed in Figure 2. The three-dimensional (3D) pattern of error distributions was restricted to be under 100% for absolute error and 500% for relative error. The corresponding two-dimensional (2D) error areas were depicted at the bottom with color bars to distinguish the different error levels. It reveals the error becomes significant with the increase of weight fraction of crystalline phases in original sample and decrease of crystallinity of internal standard (bottom right corner), which means the original amorphous phase calculation (Equation (1)) is not applicable any more. On the contrary, the data located at the top left of 2D error areas, as shown in purple and blue colors, can be accepted and applied to original equation. When the preparatory conditions were set as '$\Delta_{AE} \leq 10\% \cap \Delta_{AE} \leq 50\%$', the error-accepted zone is displayed in Figure 3. It is from the intersection operation between '$\alpha \geq 10\beta/(1 + 10\beta)$' and '$\alpha \geq 2\beta/(1 + \beta)$'. An intersection point (8/9, 4/5) can be calculated from the above equations. Therefore, α and β should satisfy the relationship of '$\alpha \geq 10\beta/(1 + 10\beta)$' when '$\beta \in (0, 80\%)$' and '$\alpha \geq 2\beta/(1 + \beta)$' when '$\beta \in (80\%, 100\%)$'.

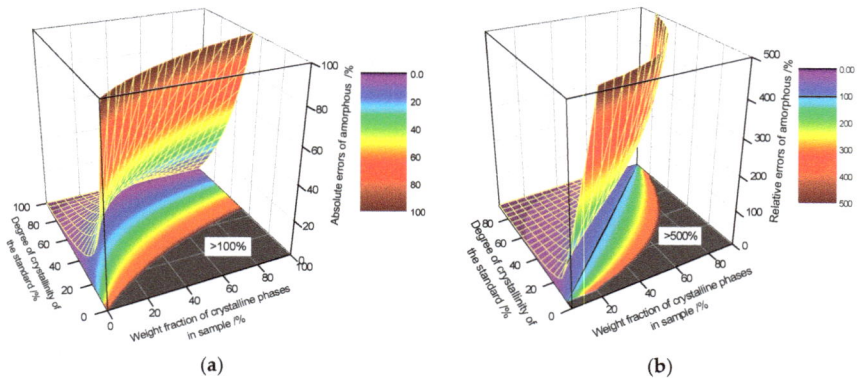

Figure 2. Error analysis between the original and improved equation (**a**: the absolute error (Δ_{AE}); **b**: the relative error (Δ_{RE})).

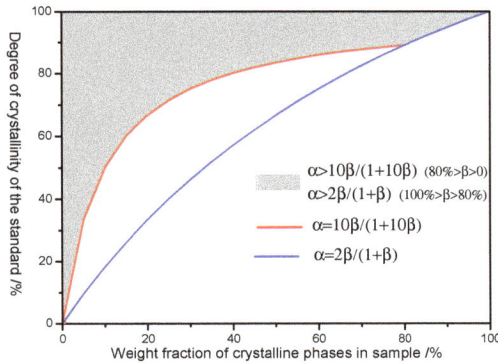

Figure 3. Accepted error-zone under the condition of '$\Delta_{AE} \leq 10\% \cap \Delta_{AE} \leq 50\%$'.

3.2. Quantitative Error Induced by Minor Standard Phase Present in Sample (SPS)

For the Rietveld/internal standard quantitative analysis, there is another problem that is often overlooked. It is also related to the standard besides the effect of its minor impurity phases above. The normal standards such as SiO_2, Al_2O_3, ZnO, and TiO_2 are preferably chosen due to their simple and known crystal structure, characteristic-sharp diffraction peaks and a variable availability of linear absorption coefficient. However, it is also important to highlight that such standards to be added for amorphous content determination may be present previously in the sample. In this work, we refer such a standard phase that is present in the sample as SPS. If SPS is minor phase and there is no better standard, they are often overlooked and assuming that contribution to quantitative accuracy is negligible. Here, we study the error analysis introduced by that extreme case and its effect on quantitative accuracy is discussed. The equations of absolute and relative errors are finally derived as Equations (9) and (10), and the intermediate derivation process was shown in the supplementary materials.

$$\Delta_{AE} = \frac{100x - y + 100}{\frac{100x}{z} + 1} \tag{9}$$

$$\Delta_{RE} = \left[\frac{100x - y + 100}{(\frac{100x}{z} + 1)y}\right] \times 100 \tag{10}$$

where x is set as '$W_{St}/100 - W_{St}$' while y and z stand for the weight fraction of amorphous phase($W^\dagger_{Amor}$) and the SPS phase ($W^\dagger_X$) in original sample, respectively.

The numerical simulation analysis of absolute and relative errors was successively performed. The results, corresponding to W_{St} and $W^\dagger_X$, were displayed in Figure 4. Figure 4a,c,e shows the absolute error distribution in which the content of amorphous phase is assumed to be 10%, 20%, and 50%, respectively. The corresponding 2D error patterns, reported by Figure 4b,d,f, displayed the relative errors distributions. It is concluded that the error level of amorphous phase quantitation is positively correlated with $W^\dagger_X$ and negatively correlated with W_{St}. The amorphous phase error becomes more sensitive with smaller of W_{St} or larger of $W^\dagger_X$. When compared with the results from 10 wt% to 50 wt% of the amorphous content, the conclusion can be drawn that quantitative error gradually dropped down at a higher dosage of internal standard. The relative errors displayed in Figure 4f are all less than 100% when the weight fraction of amorphous phase reaches 50 wt%. Likewise, the error is related with the weight fraction of standard phase present in sample and dosage of internal standard. The conditions for high quantitative accuracy of original equation are strictly satisfied with a lower amount of SPS phase, less than 2 wt%, and a higher dosage of internal standard, larger than 20 wt%. For example, if the weight percentage of SPS phase is set as 2 wt%, the absolute and relative errors are as high as 8% and 39%, respectively, when the amorphous content and dosage of internal standard are less than 20%.

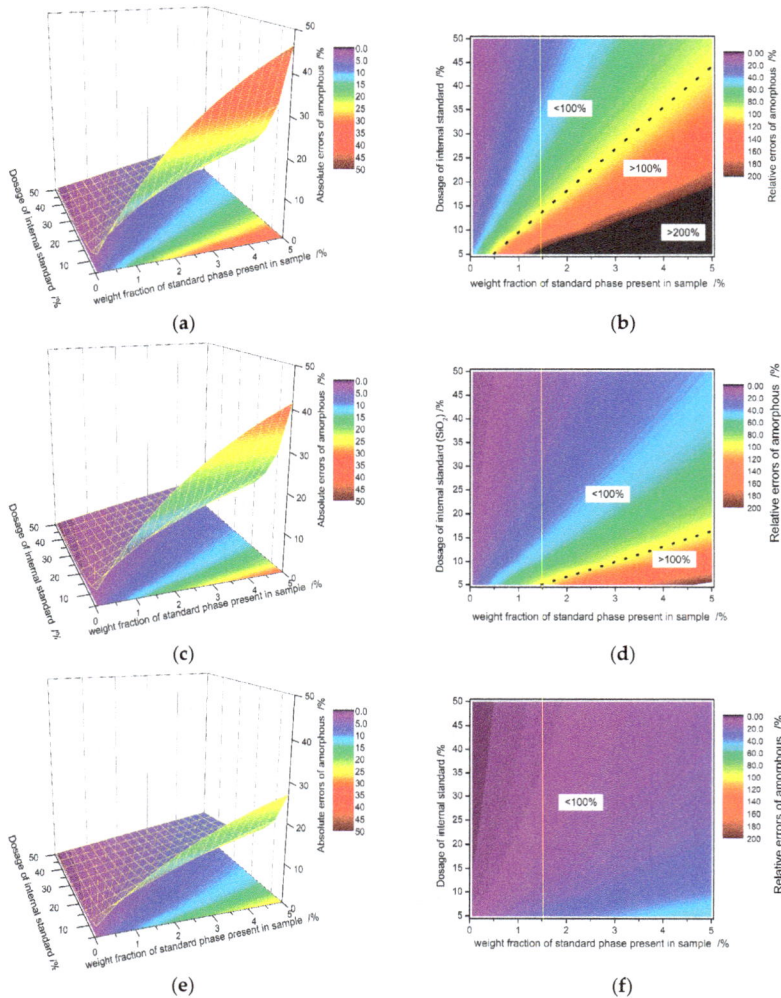

Figure 4. Absolute and relative errors analysis of amorphous phase. (**a,c,e**): the absolute error distribution in which the content of amorphous phase is assumed to be 10%, 20%, and 50%, respectively; (**b,d,f**) The corresponding 2D relative errors patterns of (**a,c,e**), respectively.

For the above quantitative errors reduction, the corrected equation for amorphous phase can be finally proposed as Equation (11), with the intermediate derivation process shown in the supplementary materials.

$$W^{\dagger}{}_{Amor} = 100 - \frac{W_{St}(R_{St} - 100)}{(R^{\dagger}{}_{X} - R_{St})(100 - W_{St})} \times 10^2 \tag{11}$$

where R_{St} and $R^{\dagger}{}_X$ stand for the Rietveld quantitative result of internal standard and the standard phase present in sample.

4. Conclusions

For Bragg-Brentano diffraction, the ignorance of the amorphous or non-quantified crystalline phases (ACn) in standard exerts significant effect on the quantitative accuracy by the Rietveld-internal standard method. The error level is only related to the crystallinity of internal standard (α) and weight percentage of amorphous phase in original sample (β), and the equations of absolute/relative errors (Δ_{AE}/Δ_{RE}) are derived as proposed. The numerical simulation analysis reveals the original equation is not applicable in most of the cases and the corresponding error-accepted zone is suggested from the relationship between α and β. When the preparatory conditions are set as '$\Delta_{AE} \leq 10\% \cap \Delta_{RE} \leq 50\%$', then the error-accepted zone can be expressed as '$\alpha \geq 10\beta/(1 + 10\beta)$', '$\beta \in (0.80\%)$' and '$\alpha \geq 2\beta/(1 + \beta)$', '$\beta \in (80\%, 100\%)$'. For the above quantitative error elimination, the equation for ACn has been improved as proposed.

Based upon Bragg-Brentano diffraction, the ignorance of minor SPS phase in original sample also has significant effect on the quantitative accuracy by the Rietveld-internal standard method. The effect is evaluated from absolute/relative errors (Δ_{AE}/Δ_{RE}) analysis with error-equations derived as proposed. The numerical simulation analysis reveals that the quantitative error has a positive relationship with the content of the SPS phase ($W^{\dagger}{}_X$) and negative relationship with internal standard (W_{St}). The error becomes more sensitive with smaller of W_{St} or larger of $W^{\dagger}{}_x$. The error-zone demonstrates that original ACn equation is inapplicable ($\Delta_{AE} > 7.8\%$, $\Delta_{RE} > 38.9\%$) under the condition of higher content of SPS phase ($W^{\dagger}{}_X > 2$wt%), while the internal standard is under 20 wt%. The improved equation for such quantitative error elimination is suggested.

Supplementary Materials: The following are available online at www.mdpi.com/2073-4352/8/3/110/s1. Part S1: the intermediate derivation process of Equation (4), Part S2: the intermediate derivation process of Equations (9) and (10), Part S3: the intermediate derivation process of Equation (11).

Acknowledgments: This Research is supported by the National Natural Science Foundation of China (No. 51602126), the National Key Research and Development Plan of China (No. 2016YFB0303505), the 111 Project of International Corporation on Advanced Cement-based Materials (No. D17001) and China and University of Jinan Postdoctoral Science Foundation funded project (No. 2017M622118, No. XBH1716). AG De la Torre thanks Spanish MINECO BIA2014-57658-C2-2-R(co-funded by FEDER) grant.

Author Contributions: Piqi Zhao, Lingchao Lu and Xin Cheng conceived and designed the experiments; Piqi Zhao performed the experiments; Piqi Zhao and Angeles G. De la Torre analyzed the data; Piqi Zhao, Xianping Liu and Lingchao Lu participated in manuscript writing and revision; All authors collaborated on the interpretation of the results and on the preparation of the manuscript.

Conflicts of Interest: The authors declare no conflict of interest.

References

1. Hull, A.W. A new method of chemical analysis. *J. Am. Chem. Soc.* **1919**, *41*, 1168–1175. [CrossRef]
2. Alexander, L.; Klug, H.P. Basic aspects of X-ray absorption in quantitative diffraction analysis of powder mixtures. *Powder Diffr.* **1948**. [CrossRef]
3. Chung, F.H. Quantitative interpretation of X-ray diffraction patterns of mixtures. II. Adiabatic principle of X-ray diffraction analysis of mixtures. *J. Appl. Crystallogr.* **1974**, *7*, 526–531. [CrossRef]
4. LeRoux, J.; Lennox, D.H.; Kay, K. Direct quantitative X-ray analysis by diffraction-absorption technique. *Anal. Chem.* **1953**, *25*, 740–743. [CrossRef]
5. Chung, F.H. Quantitative interpretation of X-ray diffraction patterns of mixtures. I. Matrix-flushing method for quantitative multicomponent analysis. *J. Appl. Crystallogr.* **2010**, *7*, 519–525. [CrossRef]
6. Zevin, L.S. A method of quantitative phase analysis without standards. *J. Appl. Crystallogr.* **1977**, *10*, 147–150. [CrossRef]
7. Popović, S.; Gržeta-Plenković, B. The doping method in quantitative X-ray diffraction phase analysis. *J. Appl. Crystallogr.* **1983**, *16*, 505–507. [CrossRef]

8. Rietveld, H.M.A. profile refinement method for nuclear and magnetic structures. *J. Appl. Crystallogr.* **2010**, *2*, 65–71. [CrossRef]

9. Bish, D.L.; Howard, S.A. Quantitative phase analysis using the Rietveld method. *J. Appl. Crystallogr.* **1988**, *21*, 86–91. [CrossRef]

10. Winburn, R.S.; Grier, D.G.; Mccarthy, G.J.; Peterson, R.B. Rietveld quantitative X-ray diffraction analysis of NIST fly ash standard reference materials. *Powder Diffr.* **2000**, *15*, 163–172. [CrossRef]

11. Gualtieri, A.; Artioli, G. Quantitative determination of chrysotile asbestos in bulk materials by combined Rietveld and RIR methods. *Powder Diffr.* **1995**, *10*, 269–277. [CrossRef]

12. Snellings, R.; Salze, A.; Scrivener, K.L. Use of X-ray diffraction to quantify amorphous supplementary cementitious materials in anhydrous and hydrated blended cements. *Cem. Concr. Res.* **2014**, *64*, 89–98. [CrossRef]

13. De la Torre, A.G.; Bruque, S.; Aranda, M.A.G. Rietveld quantitative amorphous content analysis. *J. Appl. Crystallogr.* **2001**, *34*, 196–202. [CrossRef]

14. Jansen, D.; Stabler, C.; Goetz-Neunhoeffer, F.; Dittrich, S.; Neubauer, J. Does Ordinary Portland Cement contain amorphous phase? A quantitative study using an external standard method. *Powder Diffr.* **2011**, *26*, 31–38. [CrossRef]

15. Scarlett, N.V.Y.; Madsen, I.C. Quantification of phases with partial or no known crystal structures. *Powder Diffr.* **2006**, *21*, 278–284. [CrossRef]

16. Álvarez-Pinazo, G.; Cuesta, A.M.; García-Maté, M.; Santacruz, I.; Losilla, E.R.; De la Torre, A.G.; León-Reina, L.; Aranda, M.A.G. Rietveld quantitativephaseanalysis of Yeelimite-containingcements. *Cem. Concr. Res.* **2012**, *42*, 960–971. [CrossRef]

17. Li, H.; Xu, W.; Yang, X.; Wu, J. Preparation of Portland cement with sugar filter mud as lime-based raw material. *J. Clean. Prod.* **2014**, *66*, 107–112. [CrossRef]

18. Zhao, P.; Wang, P.; Liu, X. Influence of particle size distribution on accurate X-ray quantitative analysis of tetracalcium aluminoferrite. *Adv. Cem. Res.* **2015**, *27*, 364–370. [CrossRef]

19. Madsen, I.C.; Scarlett, N.V.Y.; Kern, A. Description and survey of methodologies for the determination of amorphous content via X-ray powder diffraction. *Z. Kristallogr. Cryst. Mater.* **2011**, *12*, 944–955. [CrossRef]

20. Hermann, H.; Ermrich, M. Microabsorption Correction of X-Ray Intensities Diffracted by Multiphase Powder Specimens. *Powder Diffr.* **1989**, *4*, 189–195. [CrossRef]

21. Westphal, T.; Füllmann, T.; Pöllmann, H. Rietveld quantification of amorphous portions with an internal standard—Mathematical consequences of the experimental approach. *Powder Diffr.* **2009**, *24*, 239–243. [CrossRef]

22. Zhao, P.; Liu, X.; De la Torre, A.G.; Lu, L.; Sobolev, K. Assessment of quantitative accuracy of Rietveld/XRD analysis of the crystalline and amorphous phases in fly ash. *Anal. Method* **2017**, *9*, 2415–2424. [CrossRef]

23. De la Torre, A.G.; Aranda, M.A.G. Accuracy in Rietveld quantitative phase analysis of Portland cements. *J. Appl. Crystallogr.* **2003**, *36*, 1169–1176. [CrossRef]

24. Zhao, P.Q.; Liu, X.P.; Wu, J.G.; Wang, P. Rietveld quantification of γ-C_2S conversion rate supported by synchrotron X-ray diffraction images. *J. Zhejiang Univ. Sci. A* **2013**, *14*, 815–821. [CrossRef]

25. Guirado, F.; Galí, S. Quantitative Rietveld analysis of CAC clinker phases using synchrotron radiation. *Cem. Concr. Res.* **2006**, *36*, 2021–2032. [CrossRef]

26. León-Reina, L.; De la Torre, A.G.; Porras-Vázquez, J.M.; Cruz, M.; Ordonez, L.M.; Alcobé, X.; Gispert-Guirado, F.; Larrañaga-Varga, A.; Paul, M.; Fuellmann, T.; et al. Round robin on Rietveld quantitative phase analysis of Portland cements. *J. Appl. Crystallogr.* **2009**, *42*, 906–916. [CrossRef]

27. Larson, A.C.; Von Dreele, R.B. *General Structure Analysis System (GSAS)*; Los Alamos National Laboratory Report LAUR86-748; Los Alamos National Laboratory: Los Alamos, NM, USA, 2004.

28. Will, G.; Bellotto, M.; Parrish, W.; Hart, M. Crystal structures of quartz and magnesium germanate by profile analysis of synchrotron-radiation high-resolution powder data. *J. Appl. Crystallogr.* **1988**, *21*, 182–191. [CrossRef]

29. Kihara, K.; Donnay, G. Anharmonic thermal vibrations in ZnO. *Can. Mineral.* **1985**, *23*, 647–654.

30. Thompson, P.; Cox, D.E.; Hastings, J.B. Rietveld refinement of Debye–Scherrer synchrotron X-ray data from Al_2O_3. *J. Appl. Crystallogr.* **1987**, *20*, 79–83. [CrossRef]

31. Finger, L.W.; Cox, D.E.; Jephcoat, A.P. A correction for powder diffraction peak asymmetry due to axial divergence. *J. Appl. Crystallogr.* **1994**, *27*, 892–900. [CrossRef]

32. Brindley, G.W. The effect of grain or particle Size on X-ray reflections from mixed powders and alloys, considered in relation to the quantitative determination of crystalline substances by X-ray methods. *Philos. Mag.* **1977**, *36*, 347–369. [CrossRef]

33. Taylor, J.C.; Matulis, C.E. Absorption contrast effects in the quantitative XRD analysis of powders by full multiphase profile refinement. *J. Appl. Crystallogr.* **1991**, *24*, 14–17. [CrossRef]

© 2018 by the authors. Licensee MDPI, Basel, Switzerland. This article is an open access article distributed under the terms and conditions of the Creative Commons Attribution (CC BY) license (http://creativecommons.org/licenses/by/4.0/).

crystals

MDPI

Article

The Rietveld Refinement in the EXPO Software: A Powerful Tool at the End of the Elaborate Crystal Structure Solution Pathway

Angela Altomare [1],*, Francesco Capitelli [2], Nicola Corriero [1], Corrado Cuocci [1], Aurelia Falcicchio [1], Anna Moliterni [1] and Rosanna Rizzi [1]

[1] Institute of Crystallography-CNR, Via G. Amendola 122/o, 70126 Bari, Italy; nicola.corriero@ic.cnr.it (N.C.); corrado.cuocci@ic.cnr.it (C.C.); aurelia.falcicchio@ic.cnr.it (A.F.); annagrazia.moliterni@ic.cnr.it (A.M.); rosanna.rizzi@ic.cnr.it (R.R.)

[2] Institute of Crystallography-CNR, Via Salaria Km 29.300, 00016 Monterotondo, Italy; francesco.capitelli@mlib.ic.cnr.it

* Correspondence: angela.altomare@ic.cnr.it; Tel.: +39-080-592-9155

Received: 13 April 2018; Accepted: 2 May 2018; Published: 4 May 2018

Abstract: The Rietveld method is the most reliable and powerful tool for refining crystal structure when powder diffraction data are available. It requires that the structure model to be adjusted is as close as possible to the true structure. The Rietveld method usually represents the final step of the powder solution process, in particular when a new structure is going to be determined and published. EXPO is a software able to execute all the steps of the solution process in a mostly automatic way, by starting from the chemical formula and the experimental diffraction pattern, passing through computational methods for locating the structure model and optimizing it, and ending to the Rietveld refinement. In this contribution, we present the most recent solution strategies in EXPO, both in reciprocal and direct space, aiming at obtaining models suitable to be refined by the Rietveld method. Examples of Rietveld refinements are described, whose results are related to different solution procedures and types of compounds (organic and inorganic).

Keywords: powder structure solution; EXPO software; Rietveld refinement

---

1. Introduction

The Rietveld method [1], born from the simple and brilliant idea of refining crystal structure together with parameters describing the diffraction profile employing directly the profile intensities, has been one of the most innovative and still now widely applied methods for studying materials from powder diffraction data. It has given a great impulse to the process of crystal structure solution by powder diffraction data, expanding the fields of application of powder diffraction which, up to the end of the 1970's, was primarily used for qualitative and semiquantitative analysis. Powder diffraction without the Rietveld method would be much less popular.

The method was proposed in 1969 as best suited for neutron powder techniques, and, in particular, refining nuclear and magnetic structures, but later, it was extended to X-ray powder diagrams and used for different kinds of analysis: structural, including lattice parameters, atomic positions and occupancies, temperature vibrations (isotropic and anisotropic); quantitative phase; grain size and micro-strain (isotropic and anisotropic); stacking and twin faults. The present paper focuses on the Rietveld method for structure refinement.

The two main steps in structural analysis are solving the structure and refining the model [2–4]. The single-crystal-like procedures for structure solution and refinement require that the estimates of the structure factor moduli derived from the diffraction experiment and associated to reflections are reliable. Indeed, in the case of a single crystal, this condition is wellaccomplished and the two steps are usually carried out in a straightforward manner. In particular, the experimental moduli are efficiently used as observations in the nonlinear least-squares calculation for refining the structural parameters.

In the case of powder diffraction peak overlap, background not always simple and easily described, preferred orientation and limited experimental resolution are concurrent problems that compromise, sometimes considerably, the process of extracting the structure factor moduli from the experimental powder diffraction pattern [5], and lower the reliability of the moduli estimates. In this scenario, using the experimental structure factor moduli as observations in the least-squares minimization for refining structural parameters can be unsuccessful, or at least it needs a powerful weighting scheme of reflections for compensating for the low accuracy of the extracted moduli [6].

The Rietveld method, which uses as observations the profile intensities of the full experimental pattern, is based on the minimization of the following quantity:

$$\sum_i w_i (Y_{oi} - Y_{ci})^2 \tag{1}$$

where the summation is over the number of experimental diffraction profile intensities, w_i is a suitable weight associated to each Y_{oi} observed intensity (usually equal to $1/Y_{oi}$) and Y_{ci} is the corresponding calculated intensity. Y_{ci} is described by analytical functions, which depend on structural and profile parameters, among which we mention: atomic coordinates, occupancies, displacement factors, lattice, background, peak shape, peak width, preferred orientation, sample displacement, sample transparency, and zero-shift error. These parameters are all variables to be refined by least squares. Due to the dependence on numerous variables, the minimization process can fall into false minima or diverge. Such a risk can be reduced and possibly avoided if the quality of the experimental diffraction pattern is good, peak and background are described by suitable functions and, especially, the structure model makes physical and chemical sense. For a good outcome of the Rietveld method, it is required that the structure model to be adjusted is as close as possible to the true one.

According to (1) formula, the progress of the Rietveld minimization can be monitored by figures of merit giving the agreement between the observed and calculated intensities, among which we mention:

$$R_{wp} = \sqrt{\frac{\sum_i w_i (Y_{oi} - Y_{ci})^2}{\sum_i w_i Y_{oi}^2}} \tag{2}$$

A small R_{wp} value is an indicator of successful minimization and, if the trap of false minimum has been skipped, successful refinement and reliable final structure model.

Several computing programs have been developed for the Rietveld refinement [7]. Some of them are of general use, some are devoted to specific classes of structures (zeolites, polymers, etc.). Some packages are able to execute not only the Rietveld refinement. The software EXPO [8] belongs to this last category, being dedicated to the full pathway of structure solution by powder diffraction data. It needs only the experimental powder pattern (collected by conventional or synchrotron X-ray or neutron diffraction) and the knowledge of the chemical formula of the sample to be investigated. Supported by a high level of automatism, EXPO is able to:

- determine the cell parameters;
- identify the space group;
- solve the structure ab initio in the reciprocal space via direct methods (extract the integrated intensities from the experimental profile for estimating the observed structure factor modulus associated to each reflection, probabilistically evaluate the phases of reflections, calculate the electron density map for deriving the structure model, optimize the model);

- solve the structure in the direct space (search the best model, compatible with the expected molecular geometry, by global optimization methods;
- refine the structure model by the Rietveld method.

The EXPO graphic interface is operational to follow the evolution of the solution process, to check the results obtained in each step, and to select, if necessary, alternative nonautomatic procedures.

Recovering a structure model that makes physical and chemical sense and is close to the true one is often an arduous task whether we adopt the reciprocal space solution or the direct space option. In the first case, the main difficulty lies in the low accuracy of the extracted structure factor moduli which are actively used in the phasing process (see Section 2.1); in the second case, the difficulty is in the choice of the geometrically compatible model and in the efficiency of the optimization method (see Section 2.2).

EXPO is supported by several powerful methodological and computational procedures all aiming at providing a structure model worthy to be submitted to the last step of the solution pathway: the Rietveld refinement. In the present paper, we describe briefly the EXPO strategies and show examples of Rietveld refinement by EXPO.

2. The Crystal Structure Solution Process in EXPO

The role of the solution process which aims to obtain a complete structure model not far from the true one is relevant for the good outcome of the Rietveld refinement. After the cell parameters and space group have been recognized, the most widely used approaches for solving structure by powder diffraction data are [9]: (a) Reciprocal Space (RS) methods, in particular direct methods (they are single-crystal-like). This is the standard solution option in EXPO. The RS methods are fast in the execution time and need minimal information, but they strongly depend on the quality of the structure factor moduli extracted from the experimental pattern and on the experimental resolution. They are usually effectively combined with structure model optimization procedures which move from reciprocal to direct space and vice versa and are based on Fourier map and least-squares calculations; (b) direct space (DS) methods, also called global optimization methods. They are trial-and-error approaches which use the profile intensities, do not depend on the structure factor moduli and do not require high experimental resolution, but they need a priori knowledge of the connectivity or coordination of any of the atoms in the structure under study and, usually, are time consuming.

In recent years, the complementary features of the RS and DS methods, well supported by the availability of synchrotron light sources and fast computers, the advances in experimental devices, and the development of innovative methods and computing algorithms, have contributed to increase the number of structures successfully solved by powder data. The software package EXPO [8] is able to execute the solution process by ab initio methods and/or global optimization methods. The model generated by the solution process is then submitted to the Rietveld refinement.

If the model is resulted from the ab initio methods, even if completed and optimized by one or more of the structure model optimization strategies included in EXPO, it is usually approximate (e.g., atom positions are not very close to the true ones, rings are distorted, etc.) and, often, not fully compatible with the conditions required by the Rietveld refinement. Sometimes, information on the expected molecular geometry can be useful for restraining bond distances, angles and planes during the refinement process in order to reduce the possibility of falling into false minima.

The model derived from the global optimization methods benefits from the geometrical information on which it is based and is more suited to the Rietveld refinement, under the hypothesis that the assumed molecular geometry is not far from the true one.

A brief description of the two kinds of methods and the main features of the Rietveld refinement available in the EXPO software follow.

2.1. Solution in Reciprocal Space

Structure solution approaches working in reciprocal space, known also as traditional or ab initio or phasing methods, like direct methods (DM) and Patterson methods [10], have been firstly developed and widely applied to solve crystal structures by single-crystal data, then effectively exported to powder data.

The reciprocal space (RS) methods require minimal a priori information (i.e., the experimental powder diffraction pattern and chemical formula in addition to cell parameters and space group) and less computing time. They need the integrated intensities I of the individual reflections to be extracted from the powder diffraction pattern. Two reference approaches are devoted to this scope: the method proposed by Pawley [11] and the other from Le Bail and coworkers [12] (the latter was adopted by EXPO). The Pawley method treats the integrated intensities as variables to be refined (in addition to cell, profile and background parameters) in the least-squares that minimize the residual (1). The limit of the method is to consider I as independent observations, an assumption that can lead to refinement instability and negative intensity values. The Le Bail method is inspired by a Rietveld formula that portions the total observed intensity of a cluster of overlapping reflections proportionally to the calculated values of the single intensities. This approach reveals extremely fast and convergent but tends to the equipartition of the overall observed intensity in case of strongly overlapping reflections.

The dependence of the RS methods on the integrated intensities is a drawback, due to the unavoidable errors on the I values caused by the powder typical problems (i.e., overlap of reflections, wrong background description, preferred orientation effects, etc. see Section 1). Additional critical points of the RS approaches are related to:

(i)   *Data resolution.* Experimental diffraction data far away from the atomic resolution can prevent the success of the structure solution process or lead to a poor electron density map, and can be difficult to interpret;

(ii) *Structure complexity.* RS methods are usually able to solve successfully crystal structures with number of atoms in the asymmetric unit (N_{au}) up to about 40 [13]. $N_{au} >> 40$ is a challenging task for RS methods.

Because of the mentioned limits, the RS methods often do not provide a structure model complete and with well-located atom positions: some atoms are missed, and some others are far or very far from the true ones. Since the success of the Rietveld method is critically dependent on the completeness and accuracy of the structure model, several advanced procedures have been developed and introduced in EXPO for enhancing the power of DM and the quality of the obtained structure model. We briefly describe the basic principles of direct methods used by EXPO for obtaining the structure model and its main strategies for optimizing the model up to Rietveld method expectations.

- *Direct methods basic principles*

In the solution process by RS methods, the crystal structure is determined by calculating the electron density given by the inverse Fourier Transform of the structure factors F_h:

$$\rho(\mathbf{r}) \approx \frac{1}{V} \sum_{\mathbf{h}} |F_{\mathbf{h}}| \exp(i\varphi_{\mathbf{h}}) \exp(-2\pi i \mathbf{h} \cdot \mathbf{r}) \tag{3}$$

The diffraction experiment provides, for each $\mathbf{h}$ reflection, the integrated intensity $I_{\mathbf{h}}$ that is directly proportional to $|F_{\mathbf{h}}|^2$. The structure factor $F_{\mathbf{h}}$ is a complex quantity whose phase $\varphi_{\mathbf{h}}$ is not obtained by the diffraction experiment [10].

Direct methods are phasing methods being able to estimate $\varphi_{\mathbf{h}}$. The basic idea of DM is that the information on phases is contained on the experimental structure factor moduli, and can be derived directly (as their name suggests) from them via mathematical relationships.

The main hypotheses on which DM are based are the following:

(1)   the positivity of $\rho(\mathbf{r})$ (i.e., $\rho(\mathbf{r}) > 0$ in the unit cell);

(2)   the atomicity of $\rho(\mathbf{r})$ (i.e., $\rho(\mathbf{r})$ corresponds to discrete atoms).

These two assumptions lead to algebraic relationships involving structure factors [14,15], from which information on phases can be recovered. Wilson [16,17] opened the door to the probabilistic approach based on an additional third assumption: the atomic positions are random variables uniformly distributed on the unit cell. Since the experimental diffraction intensities are not on absolute scale and depend on the scattering angle θ, Wilson proposed a method, known as Wilson plot method, that, starting from the experimental structure factor moduli, is able to determine a scale factor K (so that the experimental $|F_\mathbf{h}|$ can be put on absolute scale) and the average isotropic displacement parameter B. Once K and B have been estimated, it is possible to calculate from $|F_\mathbf{h}|$ the normalized structure factors moduli $|E_\mathbf{h}|$ that are key magnitudes for DM. They have the great advantage to be independent of θ and correspond to an idealized point atom structure.

In the case of powder data, non-negligible errors on $I_\mathbf{h}$ can result in scarce accuracy on $|F_\mathbf{h}|$ and, consequently on $|E_\mathbf{h}|$, as well as in an unreliable phasing process and approximate result from (3).

EXPO is able to automatically carry out all the following main steps of DM:

(1)   *Normalization*: the integrated intensities are normalized by the Wilson method and the $|E_\mathbf{h}|$ values are calculated. Statistical analysis on $|E_\mathbf{h}|$ is performed in order to detect: (a) the presence or absence of an inversion center; (b) the possible presence and type of pseudotranslational symmetry [18]; (c) the preferred orientation effects [19]. The largest $|E_\mathbf{h}|$ values (i.e., $|E_\mathbf{h}| \geq 1$) reflections, the so called strong reflections, whose number is N_{larg}, are considered because they strongly contribute to the DM phasing process (see next point 2). The default number of strong reflections to be phased (N_{phas}) is automatically calculated by EXPO by taking into account the number of atoms in the asymmetric unit and the type of symmetry. N_{phas} should be at most N_{larg}; if $N_{\text{phas}} > N_{\text{larg}}$, EXPO sets $N_{\text{phas}} = N_{\text{larg}}$;

(2)   *structure invariants (s.i.) calculation*: s.i. are magnitudes that are independent of change of the origin and depend only on the structure. They are fundamental in the phasing process. For example, a s.i. of order n is the product of n normalized structure factors $E_{\mathbf{h}_1} E_{\mathbf{h}_2} \ldots \ldots E_{\mathbf{h}_n}$ with $\mathbf{h}_1 + \mathbf{h}_2 + \ldots + \mathbf{h}_n = \mathbf{0}$. In the phasing process, a special role is occupied by triplet invariants $E_\mathbf{h} E_\mathbf{k} E_{-\mathbf{h}-\mathbf{k}}$ ($n = 3$). EXPO estimates the triplet phase $\Phi_{\mathbf{h},\mathbf{k}} = \varphi_\mathbf{h} + \varphi_\mathbf{k} + \varphi_{-\mathbf{h}-\mathbf{k}}$ via the probabilistic formula $P_{10}(\Phi_{\mathbf{h},\mathbf{k}})$ [20] that depends on ten $|E|$ values. The reliability parameter $G_{\mathbf{h},\mathbf{k}}$ of the estimate of the triplet phase $\Phi_{\mathbf{h},\mathbf{k}}$ is proportional to the product $|E_\mathbf{h} E_\mathbf{k} E_{-\mathbf{h}-\mathbf{k}}|$ (it is large for strong reflections) and, in the case of a crystal structure with N non-H equal atoms in the unit cell, is inversely proportional to N: with the increasing of N, $G_{\mathbf{h},\mathbf{k}}$ becomes negligible and, consequently, the probability of DM failure increases. $G_{\mathbf{h},\mathbf{k}}$ can be positive or negative; if positive, $P_{10}(\Phi_{\mathbf{h},\mathbf{k}})$ attains its maximum at $\Phi_{\mathbf{h},\mathbf{k}} \cong 0$ (i.e., positive triplets), if negative, $P_{10}(\Phi_{\mathbf{h},\mathbf{k}})$ is maximum at $\Phi_{\mathbf{h},\mathbf{k}} \cong \pi$ (i.e., negative triplets). The positive triplets with $G_{\mathbf{h},\mathbf{k}} \geq 0.6$ are stored by EXPO and are strongly involved in the phasing process;

(3)   *Phases estimate*: a milestone for DM is the tangent formula, proposed by Karle & Hautpman [21], that derives the phases of the N_{phas} reflections starting from a subset of selected reflections (the so called starting set) and actively uses the triplets in which the reflections to be phased are embedded. The phases of the starting set can be set via a multisolution method based on magic integers [22] (this is the default choice of EXPO), or, alternatively, by a random approach starting with random phase values. DM provide several possible sets of phases that are ranked according to a suitable mathematical tool, the combined figure of merit (CFOM) [23,24], mainly based on the $|E|$ values. The largest CFOM value phase set should correspond to the correct solution. When it does not provide a plausible and chemically interpretable structure solution, EXPO offers the graphic option to conveniently explore all the generated and stored phasing sets (their number is usually 20);

(4) *Electron density map calculation*: the calculation of (3) is carried out on the largest CFOM value phase set or on some or all the DM generated and stored sets. The interpretation of $\rho(\mathbf{r})$ in terms of positions and intensities of its peaks supplies the fractional coordinates and the chemical labels of the atoms in the structure. Because of uncertainties on the experimental structure factors moduli extracted from the powder profile, the entire DM process can be unsatisfactory and the final structure model approximate. Consequently, the completion and/or optimization of the DM structure model are a fundamental request for a successful and meaningful application of the Rietveld method.

- *Model optimization*

EXPO proposes procedures, some applied in a default way and others on request when the standard approaches fail, all aiming to improve the structure model obtained by DM in such a way that the optimized structure is close to the true solution and appropriate for the Rietveld refinement. The optimization strategies are:

(1) WLSQFR (weighted least-squares Fourier recycling) [6] consists of a two-step approach alternating suitably weighted least-squares refinement (aiming to minimize the weighted squared difference between the observed and calculated intensities) and Fourier map calculations, which add missing atom positions to the refined model. The weights take into account the low accuracy of the integrated intensities estimates of the overlapping reflections and tend to prevent the domination of the refinement process by the largest intensity reflections. This optimization tool is automatically applied in a default run of EXPO in case of inorganic compounds solution.

(2) RBM (resolution bias modification). The procedure can work in direct or reciprocal space or in both spaces [25–29] and is a powerful approach able to reduce the errors on the electron density map mainly due to the limited experimental resolution. RBM is able to discard false peaks and recover the missing ones. It has revealed itself to be particularly effective for organic and metal-organic compounds (for them it is the default choice of EXPO).

(3) COVMAP [30], a procedure aiming to correct the electron density map by exploiting the principle of covariance between two points of the map, nevertheless its quality. It can locate the missing atoms by modifying the electron density map taking into account some basic crystal chemistry rules, in particular, the expected bond distances between couples of atoms.

(4) Shift_and_Fix [31], the last developed approach and effectively introduced in EXPO based on the optimization of the models derived from all the stored DM phasing sets that are automatically and sequentially processed and analyzed. Shift_and_Fix consists of two main steps, cyclically combined:

    (a) The *Shift step*, which randomly shifts a suitably chosen part of the DM structure model;

    (b) The *Fix step*, which carries out a weighted least-squares refinement of the shifted model, followed by Fourier map calculations whose coefficients are functions of the chemical content of the compound under study.

COVMAP and Shift_and_Fix are not default procedures of EXPO and can be fruitfully applied to organic, inorganic and metal-organic crystal structures.

If hydrogen atoms are present in the structure under investigation (the solution process by RS methods is not able to detect them), the EXPO graphical interface provides computational tools for positioning them geometrically (if possible) and/or detecting them by nonstandard Fourier map calculation (i.e., difference Fourier map).

2.2. Solution in Direct Space

Because of the unavoidable problems of the powder diffraction pattern (peak overlapping, background, preferred orientation), responsible of ambiguities in the integrated intensities of individual reflections, the reciprocal space methods are not always able to reach a correct solution or at least a solution satisfactory for the Rietveld refinement. Structure solution procedures alternative to the RS approaches result in the direct space (DS) methods where the fit of calculated versus experimental powder pattern is performed by moving trial molecule models inside the unit cell.

By avoiding the pattern decomposition into single integrated intensities, the DS approaches overcome the limit of the strong dependence of the RS methods on the quality of the structure factor moduli extracted from the profile and on the experimental resolution. On the other hand, the principal reason for their success is the incorporation of the prior knowledge on the expected molecular geometry of the compound under study. However, this request also represents the main limitation of the technique: the full molecular structure must be managed if the correct solution is to be obtained.

In the DS approaches, a starting model is supposed with well-characterized bond lengths and bond angles typically assumed as known and considered standard values during the solution process (this assumption is usually well placed for the organic compounds). The structural variables describing the starting model are so restricted to the external (position and orientation) plus the internal (torsion angles) degrees of freedom (DOFs) whose values cannot be determined a priori. Several random trial structures are generated from the starting model by varying DOFs describing the model location and orientation in the unit cell and its internal conformation.

The quality of each trial model is evaluated via a cost function (CF) that compares the powder diffraction pattern calculated from the current structure with the experimental one. The direct space strategy is to find the trial model corresponding to the lowest CF value, which is equivalent to exploring the hypersurface of CF(DOFs) to find the set of DOFs that define the structure and correspond to the global minimum in CF(DOFs).

Several global optimization algorithms have been adopted to attain this objective [32–34] and successfully applied for solving structures of organic, inorganic and organometallic materials. Their common limit is the computational time consumption, particularly when the number of DOFs is large. Among the DS approaches, the most widely used is the Simulated Annealing (SA) [35–39].

Two alternative DS structure solution algorithms have been implemented in EXPO: (1) a classical Simulated Annealing (default choice); (2) the most innovative GHBB-BC global optimization method [40]. They both require a starting trial structure model compatible with the expected molecular connectivity information which can be built up by using a molecular editor program or retrieved by literature in the case that a similar molecule has been already published. The hydrogen atoms (their contribution to X-ray scattering is weak) present in the starting trial model can be omitted in the DS process in order to reduce the number of DOFs and the computational time, which increases with the number of atoms. EXPO is able to read the starting model in different formats: MDL Molfile, MOPAC file, Tripos Sybyl file, Crystallographic Information File (CIF), Protein Data Bank (PDB) file, XYZ format, etc. (see the EXPO manual for more details). EXPO itself offers graphical tools for adding building blocks (tetrahedron, octahedron, square plane, cube, trigonal plane, antiprism tetragonal, prism trigonal, icosahedron, isolated atom).

The default solution process by DS methods consists of ten runs, which are automatically performed. During each run, a visual matching among observed, calculated and difference profile, together with the current trial structure model, is plotted. A plot of the CF value depending on the evolution of the process can be also monitored.

During each run, EXPO changes the torsion angles (automatically identified by the program), the orientation, and the position of the trial model in the unit cell, while bond lengths, bond angles and ring conformation are not considered as variables. For this reason, the assumed lengths and angles should match as closely as possible to those true of the studied compound if incorrect results are to be avoided. At the end of the automatic procedure, the ten solutions are ranked according to the CF value, saved in CIF files and graphically shown.

Two cost functions can be alternatively selected for driving the process towards the attainment of the best structure solution:

(1) The R_{wp} weighted profile reliability parameter (see Equation (2)) which represents the default choice;

(2) The R_I agreement factor, which compares the experimental integrated intensities $I_h(obs)$ and the intensities $I_h(calc)$ calculated by the model:

$$R_I = \sqrt{\frac{\sum\limits_h |I_h(obs) - I_h(calc)|}{\sum\limits_h I_h(obs)}} \qquad (4)$$

Even if Equation (4) requires the determination of the integrated intensities, the advantage in its use (instead of R_{wp}) is the consistent reduced computing time while, if the reflection overlapping is severe, the $I_h(obs)$ values can be unreliable, and the use of R_{wp} is preferred.

The most relevant aspects of SA and GHBB-BC in EXPO are hereinafter reported.

- *Simulated Annealing*

Simulated Annealing is an iterative metaheuristic algorithm widely used to address discrete and continuous optimization problems. Its main advantage is that during the explorative walk on the CF hypersurface via the Monte Carlo method [33,41,42], uphill moves are allowed, providing the trial model to escape from local minima in search of the global minimum. On the contrary, one shortage of SA is the high computational cost, strongly dependent above all on the chosen annealing schedule that regulates the temperature (T) parameter control. In effect, if T is reduced too rapidly, a premature convergence to a local minimum may occur; in contrast, if it is reduced too gradually, the algorithm is very slow to converge.

At the beginning of the SA process implemented in EXPO, a dialog window reporting the general settings of the procedure (i.e., cost function, experimental resolution, random seed, temperature, etc.) is shown. All the control parameters are automatically set by the program to execute an annealing schedule which represents a compromise between the requirements of maximizing the efficiency of the algorithm and minimizing the total execution time. If necessary, all the parameters can be modified by the user to better deal with the problem under study.

The SA user-friendly graphic interface of EXPO is supported by useful tools for visualizing and checking the various steps of the solution process and eventually modifying the default choices.

The dynamical occupancy correction [39] represents another important available option, which is very useful when DS methods are applied to nonmolecular crystals for which some atoms are expected to occupy special positions and different building blocks can share one or more atoms. This option can be activated graphically.

The problem of computational time required for complex structures, in general for compounds with more than 15–20 DOFs, represents a limit for the crystal structure solution by DS: a large number of SA moves per run, and a large number of runs are required to guarantee to find the global minimum and increase the frequency of correct solutions. Fortunately, this type of calculation can be easily distributed among more CPUs by a parallel version of EXPO which has been developed and is going to be improved by using the Message Passing Interface (MPI) parallelization paradigm and available for parallel machines (ordinary laptops and desktop PCs, supercomputers) with Linux operating systems.

- *GHBB-BC Method*

The recently proposed GHBB-BC algorithm has been developed for improving the features of the DS search methods as well as saving computational time. Its capabilities have been optimized for the application to compounds with a number of torsion angles lower than six and not more than two fragments in the asymmetric unit. It results from a proper combination of three DS approaches:

(1) The Big Bang-Big Crunch global optimization method (BB-BC) [43]. It is inspired by one of the cosmological theories of the universe and involves two phases: (i) the Big Bang, corresponding

to the disorder caused by the energy dissipation in which a completely random population is generated; (ii) the Big Crunch, corresponding to the order due to gravitational attraction where the population shrinks to a single good quality element represented by the centre of mass, for converging to a global optimum point.

(2) The metaheuristic Greedy Randomized Adaptive Search Procedure (GRASP) [44]. It is an iterative approach, particularly effective in finding empirically good quality solutions in a reasonable computational time for most of the real-world combinatorial optimization problems that are computationally difficult and have enormous solution spaces. Each GRASP iteration is made up of two phases: construction and local search. The construction phase progressively builds a set of feasible solutions from scratch; the local search phase investigates their neighbourhood until a local minimum is found. The best overall solution is kept as the final result.

(3) The traditional Simulated Annealing.

The GHBB-BC computational procedure starts from an expected external structure model and performs a number of possible iterations defined according to the number of structure fragments and torsions angles present in the model. The general iterative scheme can be summarized (see [40] for details): the Big Bang phase creates the initial random population whose elements are evaluated by their corresponding R_{wp} values; cycles of GRASP come after for improving the population according to an effective sample; the Big Crunch phase selects three representative population elements which are then considered as centres of mass from which a new Big Bang phase can restart; at most, three population elements are conveniently chosen and submitted to SA optimization; the global optimization attainment is achieved by picking up the best population element corresponding to the minimum R_{wp} value after it has been carefully checked; the best model is accepted and possibly used for starting with a new iteration.

To complete the framework of the mainly used powder structural solution approaches, the hybrid methods [45–48] should be considered. They result from the combination of RS and DS approaches to take advantages from the best features of the two methods. Of particular interest are two hybrid nonstandard procedures implemented in EXPO and obtained by combining Direct Methods with SA [49,50] and Monte Carlo methods [51,52], respectively.

As a final step of the solution process, the model obtained by the application of one of the DS algorithms in EXPO is then submitted to the Rietveld method. The request that it is a reasonably good model in order to succeed with the Rietveld refinement and converge to the correct solution is satisfied only if the prior information about the expected molecular geometry is in agreement with the true one. This is the reason for which the DS methods are very popular for solving organic compounds: for them, it is more difficult to fail in the building of the *prior* model. A well formulated global optimization approach is equivalent to a 'global Rietveld refinement' [53].

Compared with the RS approaches, which usually provide models so poor and far from chemically reasonable that their optimization is mandatory, the DS methods can be more profitable for both structure solution and for the Rietveld refinement (provided that the input molecular model is accurate). Critical points are: compounds with more than one chiral center and *cis/trans* isomerism, a nonplanar ring system, or an unusual combination of elements in functional groups. Experience and chemical intuition are certainly required to build the starting model.

2.3. The Rietveld Refinement

The paper by McCusker and coworkers [54] is very informative on the practical aspects of the Rietveld refinement, in particular on the usually involved parameters and strategy for their variation.

EXPO is able to carry out Rietveld refinements from X-ray or neutron powder diffraction data. Its main features are briefly described. The following parameters can be adjusted during the EXPO Rietveld execution:

(1) Parameters for correcting the systematic line-shift errors due to sample displacement, sample transparency, and zero-shift.

(2) Background parameters. The background is automatically described by empirical functions: the classical polynomial function, the Chebyshev polynomial function, and the cosine Fourier series. It can be also modeled by a mouse-click selection of points interpolated by the best fitting background curve.

(3) Parameters related to integrated intensities: scale factor and preferred orientation. Correction for the preferred orientation can be achieved by the March–Dollase function [55].

(4) Profile parameters: full width at half maximum of the peak shape and peak asymmetry. The available peak shape functions are: pseudo-Voigt, Pearson-VII, and modified Thompson–Cox–Hastings pseudo-Voigt. The correction for the peak asymmetry is applied by using the semiempirical function given in [56].

   $K\alpha_1$ and $K\alpha_2$ peak doublet, if present, can be modelled.

(5) Crystal structure parameters: lattice parameters, atomic fractional coordinates, occupation factors, and isotropic displacement parameters. Atomic displacement parameters can be refined individually or in a group of atoms with the same atomic type or the same environment.

The nonlinear least squares are carried out by employing the damped Gauss-Newton method. A backtracking line-search procedure based on cubic interpolation is used to automatically calculate the damping factor applied to the shift on parameters in order to ensure the descendant direction in each cycle of refinement and to prevent divergence [57]. The refinement convergence condition is reached when the increments on parameters become smaller with respect to their standard deviations or when the relative gradient of the χ^2 minimized function given by (1) is less than a tolerance value. Tolerance value and the maximum number of cycles can be suitably modified by the user.

The Le Bail technique [12] can be adopted to perform a full pattern decomposition prior to the Rietveld refinement in order to determine the starting values of parameters (background, peak shape, line-shift corrections and unit cell dimensions), and then subjected to refinement. This strategy is suggested especially if the available structure model is not completed [58] or when the starting model is too different from the target model.

The refinement process can be executed by following two alternative approaches: (1) the user can decide the refinement strategy via graphic interface; (2) an automatic refinement schedule can be applied. In the automatic mode, groups of parameters are refined according to a fixed sequence as established in the Rietveld guidelines [54,58]. In the last step of refinement, all parameters are refined simultaneously.

To reduce problems due to loss of experimental information and to increase the ratio 'number of observations/number of parameters to be refined', the knowledge of molecular geometry can be introduced and exploited in the refinement in the form of restraints on bond distances, angles and planes. To simplify the setting of restraints, the program is able to extract from the connectivity of the initial model the possible restraints providing a list of current and target values. The user can select the restraints to be included in the refinement procedure and eventually modify the target values.

Constraints, defined as exact mathematical relationships between least-squares parameters, can be used to reduce the number of parameters. Symmetry constraints, required to conserve the space-group symmetry rules, are mandatory and automatically deduced and imposed by EXPO. Constraints on site occupancies and on isotropic displacement parameters may be defined by the user. Hydrogen atoms can be geometrically generated in an automatic way and constrained according to the riding model approximation [59] (this strategy is usually adopted in the single-crystal refinement): H atoms are moved synchronously with the atoms to which they are bonded, thereby preserving the bond length and direction; the isotropic displacement parameters of the hydrogens are constrained to be 1.2 times that of the heavy atom to which they are attached. The factor 1.2 can be changed by graphic interface.

Several criteria of fit are available in order to monitor the progress and evaluate the quality of Rietveld refinement [54] in EXPO: the weighted profile R-factor and the unweighted profile R-factor calculated for the full pattern (R_{wp}, R_p); the corresponding background-subtracted figures (R_{wp}', R_p'); the statistically expected R value (R_{exp}); the goodness-of-fit (R_{wp}/R_{exp}); the Durbin-Watson d-statistic. R values similar to those used in the case of single-crystal data are also available: R_F and the Bragg-intensity R (R_{Bragg}) which use the structure factor moduli extracted from the experimental profile. In addition, graphical tools are available for checking: (1) the match between the observed and calculated data by visualization of the observed, calculated, and difference pattern and the cumulative χ^2 value; (2) the chemical sense of bonding and nonbonding distances, angles and displacement parameters by direct display of the crystal structure.

3. Application

3.1. Rietveld Refinement of N-Benzyl-1-(prop-2-yn-1-yl)-1H-benzo[d]imidazol-2-amine

An example of structure solution and Rietveld refinement reported here regards the organic compound *N*-Benzyl-1-(prop-2-yn-1-yl)-1H-benzo[d]imidazol-2-amine ($C_{17}H_{15}N_3$) [60] (see Figure 1 and Table 1). In this case, the X-ray powder diffraction analysis is fundamental because it gives a full structural elucidation of the products and starting materials of a new synthetic process where other approaches (NMR) could be ambiguous and error-prone. The solution was attained by the DS method, in particular by Simulated Annealing of EXPO. The trial model for starting SA was created by using the sketching facilities of ACD/ChemSketch [61] and applying the MOPAC program [62] for the geometry optimization. For the structure solution, the angular range $6° < 2\theta < 50°$ of the experimental powder diffraction profile (X-ray standard laboratory data) was used. The number of parameters, varied during the minimization process, was equal to 10: three coordinates to describe the position of the center of mass, three describing the orientation, and four torsion angles to describe the conformation. The SA algorithm, applied in a nonstandard-way, was run 20 times and the structure model corresponding to the smallest value of the cost function $R_{wp} = 7.11$ was selected. The criterion to accept the solution was based also on the soundness of the crystal packing. Then, the solution derived from the DS procedure was used as an input model for the Rietveld refinement after that the H atoms were placed in calculated positions. The peak shape was modelled by the Pearson VII function. The background was fitted by a 20 coefficients polynomial. The number of structural and profile refined variables was 112:4 cell parameters, 60 nonhydrogen atomic fractional coordinates, 20 isotropic displacement parameters, 7 profile parameters, 20 background coefficients, and 1 zero-shift parameter. Hydrogen atoms have been constrained according to the riding model approximation. The automatic refinement executed by EXPO was robust and convergence was quickly achieved yielding $R_{wp} = 2.787$ and $\chi^2 = 1.846$ (Table 1).

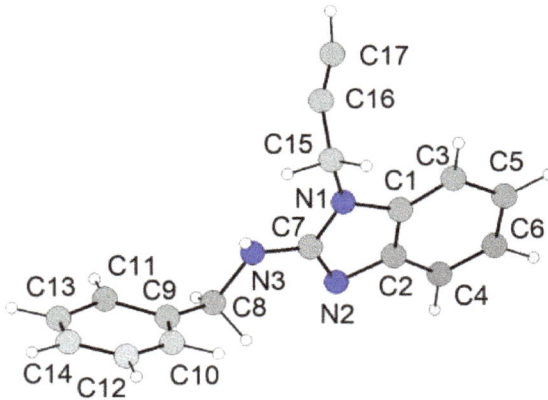

Figure 1. Crystal structure of $C_{17}H_{15}N_3$ [57].

Table 1. Crystal data, Rietveld refinement parameters and CCDC [63] depository number for $C_{17}H_{15}N_3$ [60].

Formula, formula weight	$C_{17}H_{15}N_3$, 261.32
Temperature (K), λ (Å)	293, 1.54056
System, space group	Monoclinic, $P2_1/c$
a, b, c (Å); β (°)	10.8944(2), 14.5650(2), 9.27713(15), 99.1021(10)
V (Å^3), Z	1453.53(2), 4
2θ range, step (°)	6–80, 0.02
Nr. of data points	3701
Nr. of Bragg refl., parameters	884, 112
R_p (%), R_{wp} (%), χ^2	2.015, 2.787, 1.846
R_{exp} (%), R_{Bragg} (%)	2.051, 3.298
CCDC depository nr.	CCDC1564263

3.2. Rietveld Refinement of $Ca_9RE(PO_4)_7$ (RE = La, Pr, Nd, Eu, Gd, Dy, Tm, Yb, Lu)

Another example of Rietveld refinement concerns inorganic structures, in particular the study of a set of new rare-earth tricalcium phosphates (TCP) $Ca_9RE(PO_4)_7$ (RE = La, Pr, Nd, Eu, Gd, Dy, Tm, Yb [64], and Lu [65]). TCP doped with rare earth (*RE*) elements are widely investigated because of their applications in biological imaging, owing to their strong luminescence properties [66,67]. Despite this, the available structural investigations and refinements of all the mentioned *RE* β-TCP inorganic structures from powder methods are quite lacking in literature [68,69], thus making a challenging task faced by EXPO, which worked on X-ray conventional laboratory diffraction data.

All the steps of the ab initio crystal structure solution process, from indexation to the Rietveld refinement, were performed by EXPO. The structure solution process was carried out by a default run of Direct Methods, confirming the model suggested by literature of a single-crystal study [69,70]. Detailed crystallographic information is summarized in Table 2 only for the $Ca_9Dy(PO_4)_7$ compound taken henceforth as a representative example of the series.

Table 2. Crystal data, structure refinement parameters and ICSD [70] depository number for $Ca_9Dy(PO_4)_7$ [64].

Formula, Formula weight	$Ca_9Dy(PO_4)_7$, 1187.99
Temperature (K), λ (Å)	273, 1.54056
System, space group	Rhombohedral, $R3c$
$a = b, c$ (Å)	10.4250(3), 37.301(2)
V (Å^3), Z	3510.8(3), 6
2θ range, step(°)	10–70, 0.06
Nr. of data points	1001
Nr. of Bragg refl., parameters	174, 86
R_p (%), R_{wp} (%), χ^2	2.88, 4.15, 2.06
R_{exp} (%), R_{Bragg} (%)	2.06, 7.24
ICSD depository nr.	CSD432791

In the successive Rietveld refinement, the default Pearson VII function was used for describing the peak shape. A nonstandard refinement strategy was adopted, requested by the graphic interface of EXPO: the *RE* positions were constrained to those of Ca in shared sites; the displacement parameters of P were set equal to those of O; and the sum of occupancies of Ca and *RE* elements were equal to 1.0 in shared sites with no limitation on the final charge at the site. In total, 86 parameters were refined, including the profile ones. The refinement procedure converged to low R_{exp}, R_p, and R_{wp} discrepancy indices whose values are indicative of reliable results: those for $Ca_9Dy(PO_4)_7$ are reported in Table 2 in addition to the crystal structure refinement data. The profile agreement between the observed (blue line) and the calculated pattern (red line) is shown in Figure 2. The difference pattern (violet curve) is also provided.

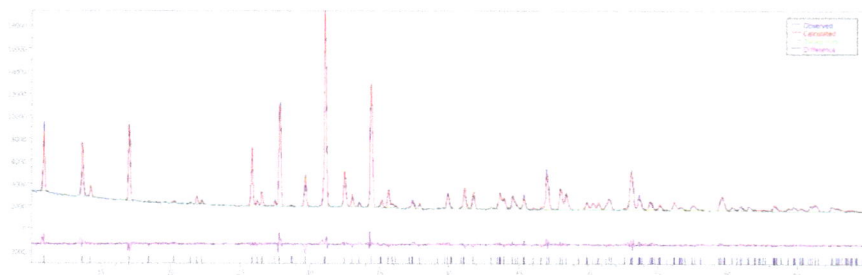

Figure 2. Rietveld plot of $Ca_9Dy(PO_4)_7$: observed diffraction profile (blue line); calculated profile (red line); difference profile (violet line); background (green) [64].

Four cationic sites are present in the structures and occupied by Ca and *RE*: three in general positions displaying eightfold coordination (named M1 and M3) and sevenfold coordination (named M2), and one on a special position displaying octahedral coordination, named M5 (Figure 3). Two of the three phosphorous and nine of the ten oxygen atoms are located on general positions, while the other atoms are located on special positions. Atom labelling was fixed according to [71].

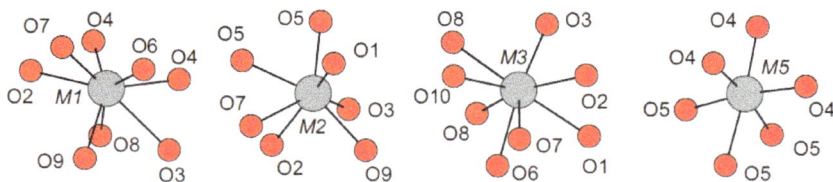

Figure 3. M1, M2, M3 and M5 cationic sites in *RE* β-TCP [64].

The aim of the Rietveld refinement is to carefully investigate the *RE* distribution within the *RE* β-TCP samples. Two different possible dopant localizations were considered: (1) *RE* in M1, M2 and M3 sites, but not in M5 according to [69]; (2) *RE* in M1, M2 and M5 sites, but not in M3. For all the compounds, the refinement of both possible configurations was tested. The monitoring of the R_{wp} values of the Rietveld refinements confirmed the cation distributions proposed in [69] for *RE* = La, Pr, Nd, Eu, Gd, and Dy behavior due to steric reasons because of the large *RE* ionic radius. Three exceptions to the described trend are represented by *RE* = Tm, Yb [64] and Lu [65], entering in site M5 and not in M3 due to their lower ionic radius for 6-coordinations, allowing them to enter in the $M5O_6$ compact octahedron. The refined site-occupancies provided a total charge very close to the ideal value of 21 valence units with minor anomalies (La and Nd the largest) within the experimental error provided by the EXPO software (see [64,65] for more details).

Considering the charge difference between RE^{3+} and Ca^{2+} and the lack of restraints on charge for mixed cationic sites, the obtained results can be considered a very reliable estimate of the *RE*/Ca ratio at each site. Rietveld refinement for $Ca_9Dy(PO_4)_7$ as well as for all $Ca_9RE(PO_4)_7$ analogue phases, was crucial for determining the exact amount of rare earth within every single calcium site, and to better understand the luminescence properties of such compounds.

4. Conclusions

The Rietveld refinement is the last necessary step in the structure solution pathway from powder diffraction data. It requires that the structure model to be adjusted is physically and chemically reliable and close to the true one. If these conditions are not fulfilled, the risk of obtaining an incorrect refined structure model is present, as well as if the minimization process has been successfully carried out and the figures of merits indicate a good fit between the observed and calculated profile. EXPO is a software package that is continuously improving the quality of the structure model obtained at the end of the structure determination, as well as the refinement process. Several strategies working both in the reciprocal and direct space can be selected with the aim of attaining a structure model suitable to the Rietveld refinement. In this way, the full solution process, from indexation to structure refinement, can be executed by EXPO, making use of default strategies or nonstandard ones.

5. Availability of EXPO

The EXPO program (last version EXPO2013) runs on any PC or workstation with the operating systems Windows, Linux or Mac OS X. For Windows, some DLLs are necessary, which are included in the distribution kit. For Linux, the binary packages for the most widely used distributions are available. The source code is also supplied and can be compiled by a Fortran95 and C++ compiler together with the GTK+2.0 and OpenGL libraries.

EXPO is available at http://www.ba.ic.cnr.it/content/expo-downloads, and the software is free for academic and nonprofit research institutions after registration. The installation instructions and the user manual are accessible via the web; documentation about the program itself and on the usage of the graphical interface is provided in HTML and PDF formats.

Author Contributions: A.A. conceived and designed the main ideas together, supervised the whole work and wrote a major part of the text; F.C. and N.C. contributed in the application session (inorganic compound); A.F. contributed in the application session (organic compound); C.C. designed and realized the study of the Rietveld refinement; A.M. designed and realized the structure study in the reciprocal space; R.R. designed and realized the structure study in direct space. All authors have read and finalized the manuscript. All authors of this paper are coauthors of the EXPO software.

Acknowledgments: The research work has been supported by the National Research Council of Italy (CNR). The costs to publish this paper in open access were covered by the Institute of Crystallography of CNR, Seat in Bari.

Conflicts of Interest: The authors declare no conflicts of interest. The founding sponsors had no role in the design of the study; in the collection, analyses, or interpretation of data; in the writing of the manuscript, and in the decision to publish the results.

References

1. Rietveld, H.M. A profile refinement method for nuclear and magnetic structures. *J. Appl. Cryst.* **1969**, *2*, 65–71. [CrossRef]

2. Pecharsky, V.K.; Zavalij, P.Y. *Fundamentals of Powder Diffraction and Structural Characterization of Materials*, 2nd ed.; Springer Science+Business Media: New York, NY, USA, 2009.

3. Clearfield, A.; Reibenspies, J.; Bhuvanesh, N. *Principles and Applications of Powder Diffraction*; Wiley: New York, NY, USA, 2008.

4. Will, G. *Powder Diffraction: The Rietveld Method and the Two Stage Method to Determine and Refine Crystal Structures from Powder Diffraction Data*; Springer: Berlin, Germany, 2010.

5. Le Bail, A. The Profile of a Bragg Reflection for Extracting Intensities. In *Powder Diffraction Theory and Practice*; Dinnebier, R.E., Billinge, S.J.L., Eds.; The Royal Society of Chemistry: Cambridge, UK, 2008; pp. 134–165.

6. Altomare, A.; Cuocci, C.; Giacovazzo, C.; Moliterni, A.G.G.; Rizzi, R. Powder diffraction: The new automatic least-squares Fourier recycling procedure in EXPO2005. *J. Appl. Cryst.* **2006**, *39*, 558–562. [CrossRef]

7. Cranswick, L.M.D. Computer Software for Powder Diffraction. In *Powder Diffraction Theory and Practice*; Dinnebier, R.E., Billinge, S.J.L., Eds.; The Royal Society of Chemistry: Cambridge, UK, 2008; pp. 494–570.

8. Altomare, A.; Cuocci, C.; Giacovazzo, C.; Moliterni, A.; Rizzi, R.; Corriero, N.; Falcicchio, A. EXPO2013: A kit of tools for phasing crystal structures from powder data. *J. Appl. Cryst.* **2013**, *46*, 1231–1235. [CrossRef]

9. Caliandro, R.; Giacovazzo, C.; Rizzi, R. Crystal Structure Determination. In *Powder Diffraction Theory and Practice*; Dinnebier, R.E., Billinge, S.J.L., Eds.; The Royal Society of Chemistry: Cambridge, UK, 2008; pp. 227–265.

10. Giacovazzo, C. *Phasing in Crystallography: A Modern Perspective*; IUCr/Oxford University Press: Oxford, UK, 2013.

11. Pawley, G.S. Unit-cell refinement from powder diffraction scans. *J. Appl. Cryst.* **1981**, *14*, 357–361. [CrossRef]

12. Le Bail, A.; Duroy, H.; Fourquet, J.L. Ab-initio structure determination of LiSbWO$_6$ by X-ray powder diffraction. *Mat. Res. Bull.* **1988**, *23*, 447–452. [CrossRef]

13. Altomare, A.; Cuocci, C.; Moliterni, A.; Rizzi, R. Single Crystal and powder XRD techniques: An overview. In *Inorganic Micro-and Nanomaterials. Synthesis and Characterization*; Dibenedetto, A., Aresta, M., Eds.; De Gruyter: Berlin, Germany, 2013; pp. 57–91.

14. Harker, D.; Kasper, J.S. Phases of Fourier coefficients directly from crystal diffraction data. *Acta Cryst.* **1948**, *1*, 70–75. [CrossRef]

15. Sayre, D. The squaring method: A new method for phase determination. *Acta Cryst.* **1952**, *5*, 60–65. [CrossRef]

16. Wilson, A.J.C. Determination of absolute from relative X-ray intensity data. *Nature* **1942**, *150*, 152. [CrossRef]

17. Wilson, A.J.C. The probability distribution of X-ray intensities. *Acta Cryst.* **1949**, *2*, 318–321. [CrossRef]

18. Cascarano, G.; Giacovazzo, C.; Luic, M. Direct methods and structures showing superstructures effects. III. A general mathematical model. *Acta Cryst.* **1988**, *A44*, 176–183. [CrossRef]

19. Altomare, A.; Cascarano, G.; Giacovazzo, C.; Guagliardi, A. Early finding of preferred orientation: A new method. *J. Appl. Cryst.* **1994**, *27*, 1045–1050. [CrossRef]

20. Cascarano, G.; Giacovazzo, C.; Camalli, M.; Spagna, R.; Burla, M.C.; Nunzi, A.; Polidori, G. The method of representations of structure seminvariants. The strengthening of triplet relationships. *Acta Cryst.* **1984**, *A40*, 278–283. [CrossRef]

21. Karle, J.; Hauptman, H. A theory of phase determination for the four types of non–centrosymmetric space groups 1P222, 2P22, 3P$_1$2, 3P$_2$2. *Acta. Cryst.* **1956**, *9*, 635–651. [CrossRef]
22. White, P.S.; Woolfson, M.M. The application of phase relationships to complex structures. VII. Magic integers. *Acta Cryst.* **1975**, *A31*, 53–56. [CrossRef]
23. Cascarano, G.; Giacovazzo, C.; Viterbo, D. Figures of merit in direct methods: A new point of view. *Acta Cryst.* **1987**, *A43*, 22–29. [CrossRef]
24. Cascarano, G.; Giacovazzo, C.; Guagliardi, A. Improved figures of merit for direct methods. *Acta Cryst.* **1992**, *A48*, 859–865. [CrossRef]
25. Altomare, A.; Cuocci, C.; Giacovazzo, C.; Kamel, G.S.; Moliterni, A.; Rizzi, R. Minimally resolution biased electron-density maps. *Acta Cryst.* **2008**, *A64*, 326–336. [CrossRef] [PubMed]
26. Altomare, A.; Cuocci, C.; Giacovazzo, C.; Moliterni, A.; Rizzi, R. Correcting resolution bias in electron density maps of organic molecules derived by direct methods from powder data. *J. Appl. Cryst.* **2008**, *41*, 592–599. [CrossRef]
27. Altomare, A.; Cuocci, C.; Giacovazzo, C.; Maggi, S.; Moliterni, A.; Rizzi, R. Correcting electron-density resolution bias in reciprocal space. *Acta Cryst.* **2009**, *A65*, 183–189. [CrossRef] [PubMed]
28. Altomare, A.; Cuocci, C.; Giacovazzo, C.; Moliterni, A.; Rizzi, R. The dual-space resolution bias correction algorithm: Application to powder data. *J. Appl. Cryst.* **2010**, *43*, 798–804. [CrossRef]
29. Altomare, A.; Cuocci, C.; Giacovazzo, C.; Moliterni, A.; Rizzi, R. The dual-space resolution bias correction in EXPO2010. *Z. Kristallogr.* **2010**, *225*, 548–551. [CrossRef]
30. Altomare, A.; Cuocci, C.; Giacovazzo, C.; Moliterni, A.; Rizzi, R. COVMAP: A new algorithm for structure model optimization in the EXPO package. *J. Appl. Cryst.* **2012**, *45*, 789–797. [CrossRef]
31. Altomare, A.; Cuocci, C.; Moliterni, A.; Rizzi, R.; Corriero, N.; Falcicchio, A. The Shift_and_Fix procedure in EXPO: Advances for solving ab initio crystal structures by powder diffraction data. *J. Appl. Cryst.* **2017**, *50*, 1812–1820. [CrossRef]
32. Hirshfeld, F.L. Symmetry in the generation of trial structures. *Acta Cryst.* **1968**, *A24*, 301–311. [CrossRef]
33. Andreev, Y.G.; Lightfoot, P.; Bruce, P.G. A General Monte Carlo Approach to Structure Solution from Powder-Diffraction Data: Application to Poly(ethylene oxide)$_3$:LiN(SO$_2$CF$_3$)$_2$. *J. Appl. Cryst.* **1997**, *30*, 294–305. [CrossRef]
34. Kariuki, B.M.; Serrano-González, H.; Johnston, R.L.; Harris, K.D.M. The application of a genetic algorithm for solving crystal structures from powder diffraction data. *Chem. Phys. Lett.* **1997**, *280*, 189–195. [CrossRef]
35. Kirkpatrick, S. Optimization by simulated annealing: Quantitative studies. *J. Stat. Phys.* **1984**, *34*, 975–986. [CrossRef]
36. David, W.I.F.; Shankland, K.; van de Streek, J.; Pidcock, E.; Motherwell, W.D.S.; Cole, J.C. DASH: A program for crystal structure determination from powder diffraction data. *J. Appl. Cryst.* **2006**, *39*, 910–915. [CrossRef]
37. Engel, G.E.; Wilke, S.; König, O.; Harris, K.D.M.; Leusen, F.J.J. PowderSolve—A complete package for crystal structure solution from powder diffraction patterns. *J. Appl. Cryst.* **1999**, *32*, 1169–1179. [CrossRef]
38. Coelho, A.A. TOPAS and TOPAS-Academic: An optimization program integrating computer algebra and crystallographic objects written in C++. *J. Appl. Cryst.* **2018**, *51*, 210–218. [CrossRef]
39. Favre-Nicolin, V.; Černỳ, R. FOX, 'free objects for crystallography': A modular approach to ab initio structure determination from powder diffraction. *J. Appl. Cryst.* **2002**, *35*, 734–743. [CrossRef]
40. Altomare, A.; Corriero, N.; Cuocci, C.; Falcicchio, A.; Moliterni, A.; Rizzi, R. Direct space solution in the EXPO package: The combination of the HBB-BC algorithm with GRASP. *J. Appl. Cryst.* **2018**, *51*, 505–513. [CrossRef]
41. Harris, K.D.M.; Tremayne, M.; Lightfoot, P.; Bruce, P.G. Crystal Structure Determination from Powder diffraction data by Monte Carlo Methods. *J. Am. Chem. Soc.* **1994**, *116*, 3543–3547. [CrossRef]
42. Tremayne, M.; Kariuki, B.M.; Harris, K.D.M.; Shankland, K.; Knigh, K.S. Crystal Structure Solution from Neutron Powder Diffraction Data by a New Monte Carlo Approach Incorporating Restrained Relaxation of the Molecular Geometry. *J. Appl. Cryst.* **1997**, *30*, 968–974. [CrossRef]
43. Erol, O.K.; Eksin, I. A new optimization method: Big Bang-Big Crunch. *Adv. Eng. Softw.* **2006**, *37*, 106–111. [CrossRef]
44. Feo, T.A.; Resende, M.G.C. Greedy randomized adaptive search procedures. *J. Glob. Optim.* **1995**, *6*, 109–113. [CrossRef]

45. Johnston, J.C.; David, W.I.F.; Markvardsen, A.J.; Shankland, K. A hybrid Monte Carlo method for crystal structure determination from powder diffraction data. *Acta Cryst.* **2002**, *A58*, 441–447. [CrossRef]

46. Brenner, S.; McCusker, L.B.; Baerlocher, C. Using a structure envelope to facilitate structure solution from powder diffraction data. *J. Appl. Cryst.* **1997**, *30*, 1167–1172. [CrossRef]

47. Wu, J.; Leinenweber, K.; Spence, J.C.H.; O'Keeffe, M. Ab initio phasing of X-ray powder diffraction patterns by charge flipping. *Nature Mater.* **2006**, *5*, 647–652. [CrossRef] [PubMed]

48. Baerlocher, C.; McCusker, L.B.; Palatinus, L. Charge flipping combined with histogram matching to solve complex crystal structures from powder diffraction data. *Z. Krystallogr.* **2007**, *222*, 47–53. [CrossRef]

49. Altomare, A.; Caliandro, R.; Giacovazzo, C.; Moliterni, A.G.G.; Rizzi, R. Solution of organic crystal structures from powder diffraction by combining simulated annealing and direct methods. *J. Appl. Cryst.* **2003**, *36*, 230–238. [CrossRef]

50. Altomare, A.; Caliandro, R.; Cuocci, C.; Giacovazzo, C.; Moliterni, A.G.G.; Rizzi, R.; Platteau, C. Direct methods and simulated annealing: A hybrid approach for powder diffraction data. *J. Appl. Cryst.* **2008**, *41*, 56–61. [CrossRef]

51. Giacovazzo, C.; Altomare, A.; Cuocci, C.; Moliterni, A.G.G.; Rizzi, R. Completion of crystal structure by powder diffraction data: A new method for locating atoms with polyhedral coordination. *J. Appl. Cryst.* **2002**, *35*, 422–429. [CrossRef]

52. Altomare, A.; Giacovazzo, C.; Guagliardi, A.; Moliterni, A.G.G.; Rizzi, R. Completion of crystal structures from powder data: The use of the coordination polyhedra. *J. Appl. Cryst.* **2000**, *33*, 1305–1310. [CrossRef]

53. Shankland, K. Global Rietveld refinement. *J. Res. Natl. Inst. Stand. Technol.* **2004**, *109*, 143–154. [CrossRef] [PubMed]

54. McCusker, L.B.; Von Dreele, R.B.; Cox, D.E.; Louër, D.; Scardi, P. Rietveld refinement guidelines. *J. Appl. Cryst.* **1999**, *32*, 36–50. [CrossRef]

55. Dollase, W.A. Correction of intensities for preferred orientation in powder diffractometry: Application of the March model. *J. Appl. Cryst.* **1986**, *19*, 267–272. [CrossRef]

56. Bérar, J.-F.; Baldinozzi, G. Modeling of line-shape asymmetry in powder diffraction. *J. Appl. Cryst.* **1993**, *26*, 128–129. [CrossRef]

57. Dennis, J.E., Jr.; Schnabel, R.B. Nonlinear Least Squares. In *Numerical Methods for Unconstrained Optimization and Nonlinear Equations*; Society for Industrial and Applied Mathematics (SIAM): Philadelphia, PA, USA, 1996; pp. 218–238.

58. Young, R.A. Introduction to the Rietveld method. In *The Rietveld Method*; Young, R.A., Ed.; Oxford University Press: New York, NY, USA, 1996; pp. 32–36.

59. Müller, P.; Herbst-Irmer, R.; Schneider, T.R.; Sawaya, M.R. Hydrogen atoms. In *Crystal Structure Refinement: A Crystallographer's Guide to SHELXL*; Müller, P., Ed.; Oxford University Press: New York, NY, USA, 2006; pp. 29–31.

60. Mancuso, R.; Veltri, L.; Russo, P.; Grasso, G.; Cuocci, C.; Romeo, R.; Gabriele, B. Palladium-Catalyzed Carbonylative Synthesis of Functionalized Benzimidazopyrimidinones. *Synthesis* **2018**, *50*, 267–277. [CrossRef]

61. ACD/ChemSketch. Available online: http://www.acdlabs.com/resources/freeware/chemsketch/ (accessed on 20 July 2017).

62. Stewart, J.J.P. *MOPAC2016*; Stewart Computational Chemistry: Springs, CO, USA, 2016.

63. Groom, C.R.; Bruno, I.J.; Lightfoot, M.P.; Ward, S.C. The Cambridge Structural Database. *Acta Cryst.* **2016**, *B72*, 171–179. [CrossRef] [PubMed]

64. El Khouri, A.; Elaatmani, M.; Della Ventura, G.; Sodo, A.; Rizzi, R.; Rossi, M.; Capitelli, F. Synthesis, structure refinement and vibrational spectroscopy of new rare-earth tricalcium phosphates $Ca_9RE(PO_4)_7$ (RE = La, Pr, Nd, Eu, Gd, Dy, Tm, Yb). *Ceram. Int.* **2017**, *43*, 15645–15653. [CrossRef]

65. Capitelli, F.; Rossi, M.; El Khouri, A.; Elaatmani, M.; Corriero, N.; Sodo, A.; Della Ventura, G. Synthesis, structural model and vibrational spectroscopy of lutetium tricalcium phosphate $Ca_9Lu(PO_4)_7$. *J. Rare Earths* **2018**, in press.

66. Madhukumar, K.; Varma, H.K.; Komath, M.; Elias, T.S.; Padmanabhan, V.; Nair, M.K. Photoluminescence and thermoluminescence properties of tricalcium phosphate phosphors doped with dysprosium and europium. *Bull. Mater. Sci.* **2007**, *30*, 527–534. [CrossRef]

67. Dorozhkin, S.V. Calcium Orthophosphates bioceramics. *Ceram. Int.* **2015**, *41*, 13913–13966. [CrossRef]

68. Lazoryak, B.I.; Kotov, R.N.; Khasanov, S.S. Crystal structure of $Ca_{19}Ce(PO_4)_{14}$. *Russ. J. Inorg. Chem.* **1996**, *41*, 1225–1228.

69. Ait Benhamou, R.; Bessière, A.; Wallez, G.; Viana, B.; Elaatmani, M.; Daoud, M.; Zegzouti, A. New insight in the structure–luminescence relationships of $Ca_9Eu(PO_4)_7$. *J. Sol. State Chem.* **2009**, *182*, 2319–2325. [CrossRef]

70. *Inorganic Crystal Structure Database (ICSD)*; Version 2017-2; Fachinformationszentrum: Karlsruhe, Germany, 2017.

71. Yashima, M.; Sakai, A.; Kamiyama, T.; Hoshikawa, A. Crystal structure analysis of β-tricalcium phosphate $Ca_3(PO_4)_2$ by neutron powder diffraction. *J. Sol. State Chem.* **2003**, *175*, 272–277. [CrossRef]

© 2018 by the authors. Licensee MDPI, Basel, Switzerland. This article is an open access article distributed under the terms and conditions of the Creative Commons Attribution (CC BY) license (http://creativecommons.org/licenses/by/4.0/).

crystals

MDPI

Article

The Effect of Skelp Thickness on Precipitate Size and Morphology for X70 Microalloyed Steel Using Rietveld Refinement (Quantitative X-ray Diffraction)

Corentin Chatelier [1,2,†], J. Barry Wiskel [1,*], Douglas G. Ivey [1] and Hani Henein [1]

[1] Department of Chemical and Materials Engineering, University of Alberta, Edmonton, AB T6G 2V4, Canada; chatelie@ualberta.ca or corentin.chatelier@univ-lorraine.fr (C.C.); divey@ualberta.ca (D.G.I.); henein4876@killamtrusts.ca (H.H.)

[2] Mines Nancy, Université de Lorraine, Campus Artem, 92 rue Sergent Blandan, F-54000 Nancy, France

[*] Correspondence: bwiskel@ualberta.ca; Tel.: +1-780-492-6178; Fax: +1-780-492-2881

[†] Current address: Institut Jean Lamour, Université de Lorraine F-54000 Nancy, France and Synchrotron SOLEIL, F-91190 Saint-Aubin, France.

Received: 11 June 2018; Accepted: 29 June 2018; Published: 12 July 2018

Abstract: Precipitates in thin-walled (11 mm) and thick-walled X70 (17 mm) microalloyed X70 pipe steel are characterized using Rietveld refinement (a.k.a. quantitative X-ray diffraction (QXRD)), inductively coupled plasma mass spectrometry (ICP), and energy-dispersive X-ray spectroscopy (EDX) analyses. Rietveld refinement is done to quantify the relative abundance, compositions, and size distribution of the precipitates. EDX and ICP analyses are undertaken to confirm Rietveld refinement analysis. The volume fraction of large precipitates (1 to 4 μm—mainly TiN rich precipitates) is determined to be twice as high in the thick-walled X70 steel (0.07%). Nano-sized precipitates (<20 nm) in the thin-walled steel exhibit a higher volume fraction (0.113%) than in the thick-walled steel (0.064%). The compositions of the nano-sized precipitates are similar for both steels.

Keywords: Rietveld refinement; quantitative X-ray diffraction; microalloyed steels; matrix dissolution; X70; ICP

1. Introduction

Microalloyed steels are a type of high strength, low alloy steel containing additions of carbon (C), nitrogen (N), niobium (Nb), titanium (Ti), and/or vanadium (V) in amounts less than 0.1 wt %. These steels may also contain molybdenum (Mo) and chromium (Cr) in amounts exceeding 0.1 wt %. Microalloyed steels are widely used in the pipeline industry [1,2] due to their good strength, weldability, and toughness. Strengthening is achieved primarily via grain size reduction, but precipitation of second phase particles can also contribute to the strength of the steel [3–5]. Precipitation strengthening can be calculated by the following equation [6]:

$$\sigma_{ppt}(\text{MPa}) = \left(\frac{10.8 V_f^{1/2}}{X} \right) \ln \left(\frac{X}{6.125 \times 10^{-4}} \right) \tag{1}$$

where V_f is the volume fraction of precipitates and X is the mean diameter (μm) of the precipitates. Changing the steel composition and/or processing conditions affects the value of both V_f and X and, ultimately, the mechanical properties of the steel. Therefore, the ability to quantify precipitate characteristics is important.

The volume fraction of precipitates in microalloyed steels is very low ($\approx$0.1%), and the size of some of the precipitates can be very small (less than 10 nm). Conventional characterization techniques, such as scanning electron microscopy (SEM) [7] or transmission electron microscopy (TEM) [3], can be used, but it is difficult and/or time consuming to obtain statistically relevant information. The size distribution of precipitates and volume fraction of nano-sized precipitates can also be determined using small angle neutron scattering (SANS) [8–10]. However, neither the chemical composition nor the morphology can be easily determined using SANS. A supplemental precipitate quantification technique, such as Rietveld refinement of X-ray diffraction (XRD) patterns, is needed to properly characterize precipitates in microalloyed steels.

To analyze precipitates using Rietveld refinement, it is necessary to extract the precipitates from the steel using a matrix dissolution technique [11,12]. Precipitate extraction allows for the analysis of a statistically significant number of precipitates. In addition, the extracted precipitates can be characterized using other analytical techniques, such as SEM coupled with energy dispersive X-ray (EDX) analysis.

The Rietveld method, in conjunction with XRD, which is commonly referred to as quantitative XRD (QXRD), is a characterization methodology that calculates a theoretical XRD pattern and matches it with the experimental diffraction pattern [13,14]. A number of different parameters are used to calculate the theoretical pattern including the lattice parameter(s) of each phase, the atomic composition(s), the crystallite size (L_{vol} in nm), the lattice strain (ε_0), and the relative abundance (weight and/or volume fraction) of precipitates. The value of L_{vol} is related to the mean precipitate radius (R) by the following equation [15]:

$$L_{vol} = \frac{3 \cdot R \cdot (1 + c)^3}{2} \tag{2}$$

where $c = 0$ for a monodisperse spherical distribution and $c = 0.2$ for a typical lognormal distribution [16]. A lognormal distribution ($c = 0.2$) was assumed during precipitate analysis. QXRD has been successfully applied to precipitate analysis in Grade 100 steel [12].

2. Materials and Methods

2.1. Steels Analyzed

Conventional 11 mm thick X70 steel (X70) and a 17 mm thick X70 steel (TWX70) skelp were analyzed in this work. The mechanical properties (yield strength, ultimate tensile strength, and percentage of elongation) and composition of each steel are shown in Table 1. The X70 and TWX70 steels have very similar processing conditions (similar finish rolling temperature ($\approx$800 °C) and runout table cooling rates ($\approx$15 °C/s)). The primary difference between the two steels is the thickness and variations in the C, Nb, Mo, and Cr content. The skelp thickness does not markedly affect the mechanical properties of the steels (less than 5% difference for the yield and ultimate tensile strength). Both steels are within the specifications of X70 steels.

Table 1. Composition of X70 and TWX70 steels.

Steel	Thickness	YS	UTS	%El	C	N	Nb	Mo	Ti	Cr	V
	(mm)	(MPa)	(MPa)		(wt %)	ppm	(wt %)	(wt %)	(wt %)	(wt %)	(wt %)
X70	11	524.0	651.6	37	0.052	70	0.09	0.13	0.016	0.23	0.003
TWX70	17	554.3	679.1	38	0.043	90	0.07	0.19	0.016	0.09	0.003

2.2. Matrix Dissolution and Precipitate Collection

A mixture of 6N HCl and distilled water, according to ASTM standard E194-90 [17], was used to dissolve samples of each steel. The samples were taken from the full thickness of the skelp. Table 2 summarizes the initial weight of each sample, the volume of solution used, the dissolution temperature, the time for dissolution, and the weight of residue collected.

Table 2. Dissolution parameters for X70 and TWX70 steels.

Steel	Weight Steel	Volume Solution	T	Time	Weight Residue
	(g)	(mL)	(°C)	(weeks)	(mg)
X70	48.1474	2500	90	3	67.1
TWX70	10.1308	500	90	1	12.2

Lu [3] observed the presence of amorphous SiO_2 in the residues using a similar dissolution technique to that described in this work. The amorphous phase presented difficulties in subsequent SEM/TEM and QXRD analyses. To remove dissolved O_2 from the dissolution system (and prevent the formation of the silica), N_2 was directly injected into the solution before starting and during the dissolution process. This procedure reduced the amount of amorphous SiO_2 collected in the residues [18]. Following complete dissolution, the solution was centrifuged (to separate the liquid from the solid residues) using a Sorvall RC-6 centrifuge (Mandel) operating at a speed of 18,300 RPM. The residue free liquid was then analyzed using inductively coupled plasma mass spectrometry (ICP), and the solid residues were analyzed using XRD and EDX analysis in the SEM.

2.3. ICP Analysis of the Supernatant Solution

Inductively coupled plasma (ICP) mass spectroscopy (Perkin Elmer Elan 6000 ICP-MS, Waltham, MA, USA) was used to measure the concentration of alloying elements (Ti, Nb, Mo, Cr, etc.) that remained (i.e., were not present as precipitates) in the dissolution solution. Calibration of the ICP system was undertaken using a four point calibration method [18]. Based on the mass of the steel sample and the volume of the initial solution, the mass fraction of each element dissolved in the solution was calculated. This information was used to complete the mass balance for the quantitative XRD analysis of the collected precipitates and will be discussed in a subsequent section.

2.4. XRD of the Solid Residue

A Rigaku Ultima IV diffractometer was used to obtain diffraction patterns from the residues. As the amount of residue collected was quite low ($\approx$5–10 mg per 10 g of steel), a quartz sample holder was used to minimize diffraction issues related to sample thickness. Table 3 provides the X-ray diffraction parameters used. A LaB_6 standard was used to calibrate instrument broadening.

Table 3. X-ray diffraction parameters.

Parameter	Value
Radiation	Cobalt
Detector	D/Tex with Fe filter
2θ range	5–100°
Scan	0.02°/step
Scan speed	2°/min

2.5. Rietveld Refinement and Analysis

Rietveld refinement was undertaken for each diffraction pattern using TOPAS Academic Software [19] (4.1, Bruker AXS Inc., Madison, WI, USA, 2007). This software uses a linear least squares method to predict the measured X-ray diffraction pattern. The parameters used in the fitting calculation include the lattice parameters for each phase, the occupancy of each atom in the unit cell, the scale factor (relative intensity of each phase), and the effects of crystallite size (L_{vol}) and strain. The goodness of the fit was evaluated using the lower weighted profile R-factor, R_{wp} [20].

The relative weight fraction (W_j) of each phase (i.e., precipitate type) observed in the diffraction pattern was calculated using Equation (3) [21].

$$W_j = \frac{S_j \cdot M_j \cdot Z_j \cdot V_j}{\sum\limits_{i=1}^{n} S_i \cdot M_i \cdot Z_i \cdot Vi} \tag{3}$$

where S is the scale factor calculated by TOPAS, n is the number of phases, M is the mass of the unit cell, Z is the number of atoms in the unit cell, and V is the unit cell volume of the phase. Using the relative weight fraction and atomic occupancy of each phase, it was possible to calculate the total amount of each element in all the precipitate types. The mass fraction of each element that remained in solution (i.e., the elements not in precipitate form) was then calculated and compared to the ICP results. From the value of crystallite size (L_{vol}), the mean radius R of a particular precipitate type was calculated using Equation (2).

2.6. SEM-EDX Analysis of Collected Precipitates

A Zeiss FE-SEM was used to image the precipitates collected in the residue. A Bruker EDX system was utilized to analyze the composition of each precipitate. In order to improve imaging of the fine precipitates, a low accelerating voltage (between 5 and 10 kV) was used. Samples were coated with a thin layer of carbon to minimize charging. For the composition analysis, an accelerating voltage of 20 kV was used.

A major issue with SEM-EDX analysis was that the fine precipitates tended to agglomerate on drying. A suspension of ethanol and residue was prepared using ultrasonic vibration. A small amount of the suspension was place on carbon tape. Once the ethanol evaporated, the individual precipitates remaining on the tape were analyzed. In total, using this technique, 82 nano-sized precipitates were analyzed for the X70 steel and 112 nano-sized precipitates were analyzed for the TWX70 steel [18].

3. Results

3.1. ICP Results

Based on the nominal composition (Table 1), the initial weight of the sample, and the mass fraction of each element dissolved in the solution determined by ICP, the wt % of each element present in solid solution in the original steel sample or in precipitate form was calculated. Figure 1 graphically illustrates the amount of each element either in solid solution or in precipitate form for both X70 and TWX70. For the X70 steel (Figure 1a), the amounts of Nb and Ti remaining in solution are relatively low (compared with the nominal composition). Conversely, most of the Cr and V remain in solution. A small amount of the Mo is in precipitate form, but the majority remains in solution. For the TWX70 steel (Figure 1b), the amounts of Nb and Ti remaining in solution are relatively low (compared with the nominal composition), and most of the Cr and V remains in solution. As with the X70 steel, a small amount of the Mo is in precipitate form, but the majority remains in solution.

3.2. SEM Size Analysis

Figure 2 shows an SEM image of individual precipitates (circled in green) that were retained on the carbon tape following evaporation of the ethanol and precipitate suspension described in Section 2.6. Small agglomerates of precipitates (circled in red) are also observed. For the purpose of this work, only the size of the individual precipitates was measured.

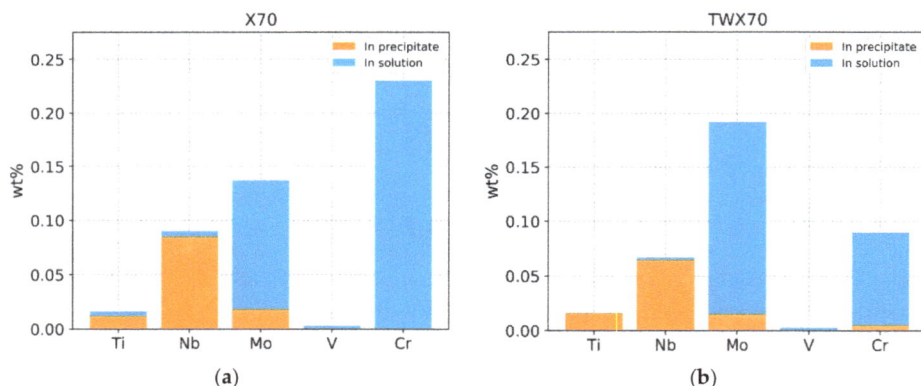

Figure 1. Composition in (wt %) of each element either in solid solution or in precipitate form for (**a**) X70 steel and (**b**) TWX70.

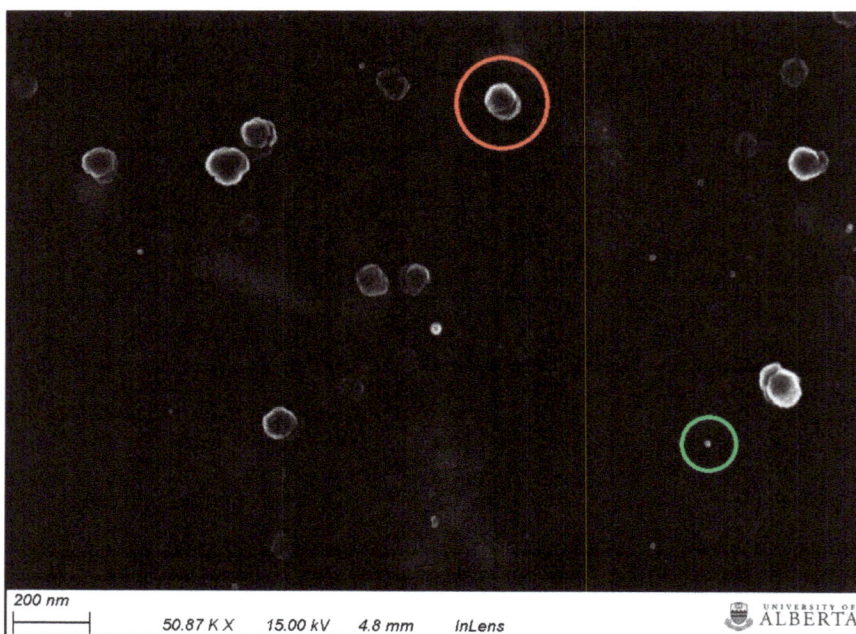

Figure 2. SEM secondary electron (SE) image of precipitates retained on carbon tape for X70 steel. An individual particle is circled in green while an agglomerate is circled in red.

Figure 3a shows the measured size distribution of precipitates extracted from the X70 steel and Figure 3b shows the size distribution of precipitates from the TWX70 steel. For the X70 steel, the average size of the fine precipitates is 7.95 nm with a standard deviation of 2.19 nm. The maximum size measured was 15.2 nm and the minimum size was 3.6 nm. For the TWX70 steel, two overlapping distributions were observed. The average size of the finer distribution was 4.4 nm and the average size of the larger precipitate distribution was 10 nm, with a maximum size measured of 13.9 nm. The overall average size was 9.15 nm.

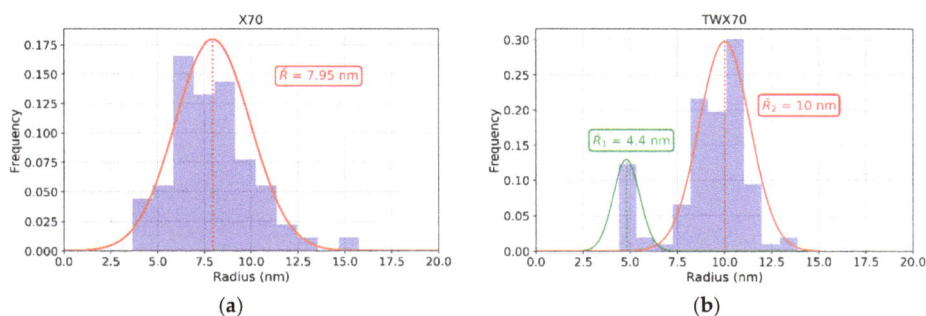

Figure 3. Measured size distributions of nano-sized precipitates from SEM images for (**a**) X70 steel (82 particles) and (**b**) TWX70 steel (112 particles).

3.3. Composition Analysis

The composition of individual precipitates was determined using SEM-EDX analysis. Figure 4 shows the atomic fraction of Nb, Ti, and Mo measured for both fine and large (μm scale) particles for the X70 steel. The large precipitates are Ti-rich, while the fine (nano) precipitates are composed primarily of Nb with some Mo and Ti. For both precipitate sizes, Nb replaces Ti in the precipitate structure. The atomic fraction ranges from 0.95 to 0.75 Ti for the large precipitates, while the fine precipitates exhibit a Nb atomic fraction range from 0.55 to 0.93, with the majority >0.7 Nb. The Mo present in the fine precipitates is relatively constant at an atomic fraction of ≈0.05 and is relatively independent of the Ti atomic fraction. The Mo content of the large precipitates is virtually nil.

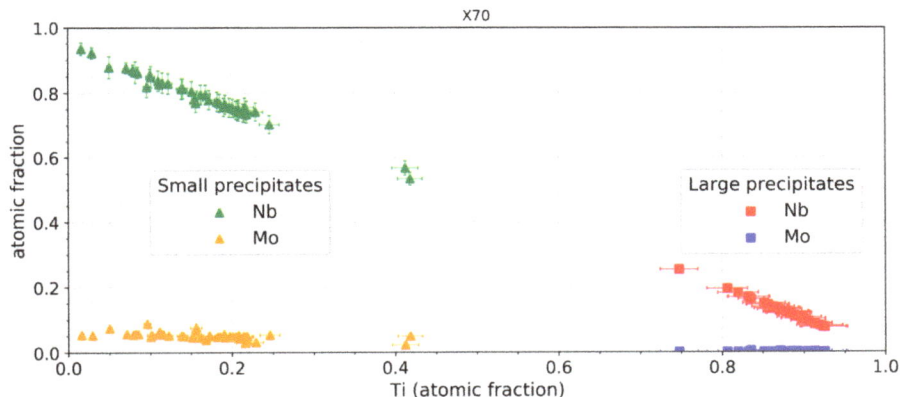

Figure 4. Atomic fraction of X70 steel precipitates measured using EDX analysis.

Figure 5 shows the atomic fraction of Nb, Ti, and Mo, measured for both fine and large (μm scale) precipitates for TWX70 steel. As observed for the X70 steel, the large precipitates are Ti-rich, while the fine (nano) precipitates are composed primarily of Nb, with lower levels of Mo and Ti. The atomic fraction of the large precipitates ranges from 0.95 to 0.75 Ti. The fine precipitates exhibit a Nb atomic fraction range from 0.4 to 0.83; however, unlike the X70 steel, the nano-sized precipitates exhibit two groupings of composition. In one grouping, there is range of Nb atomic fractions (0.4 to 0.8) with Ti content decreasing as Nb content increases, and the Mo content is relatively constant at an atomic fraction of ≈0.10. In the second grouping (circled in Figure 6), the Mo content is significantly higher, i.e., with an atomic fraction ranging from 0.22 to 0.28, accompanied by a lower Nb content

(0.52 to 0.56 atomic fraction). The presence of the second grouping of precipitates is attributed to the higher Mo content and lower Nb content of the TWX70 steel (Table 1).

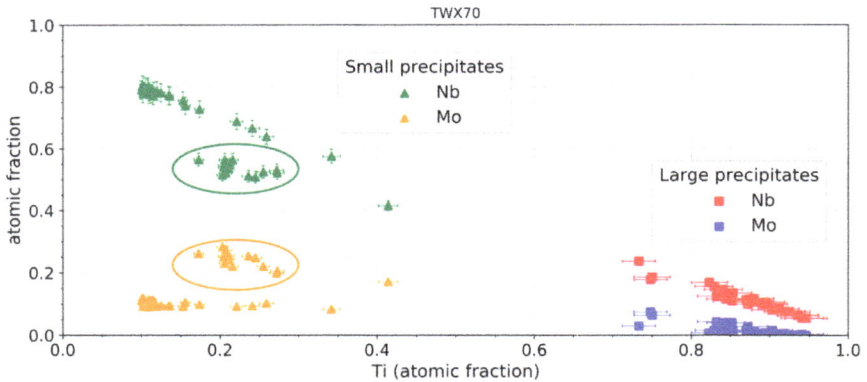

Figure 5. Atomic fraction for TWX70 steel precipitates measured using EDX analysis.

3.4. X-ray Diffraction

An XRD pattern of the residue collected from the X70 steel in shown in Figure 6. The relatively small and broad peak at $2\theta \approx 27°$ is attributed to a small amount of amorphous SiO_2. Couplets of peaks (a large intensity peak followed by a significantly less intense diffraction peak) are observed throughout the diffraction pattern. The large first peak is associated with a Nb/C-rich precipitates, whereas the second (and smaller) peak in each set is associated with Ti/N-rich precipitates. Both peaks in each couplet are characteristic of a NaCl-type structure (Fm3m space group). The diffraction planes (e.g., (111)) for each couplet are included in Figure 6.

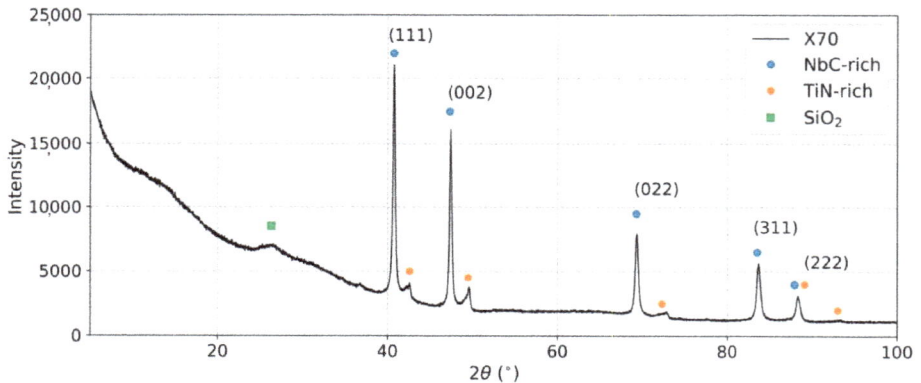

Figure 6. XRD pattern from X70 steel residue.

Figure 7 shows the measured XRD pattern of the collected residue for TWX70 steel. As with the X70 steel, couplets of diffraction peaks are observed throughout the diffraction pattern and are associated with Nb/C-rich precipitates and Ti/N-rich precipitates, respectively.

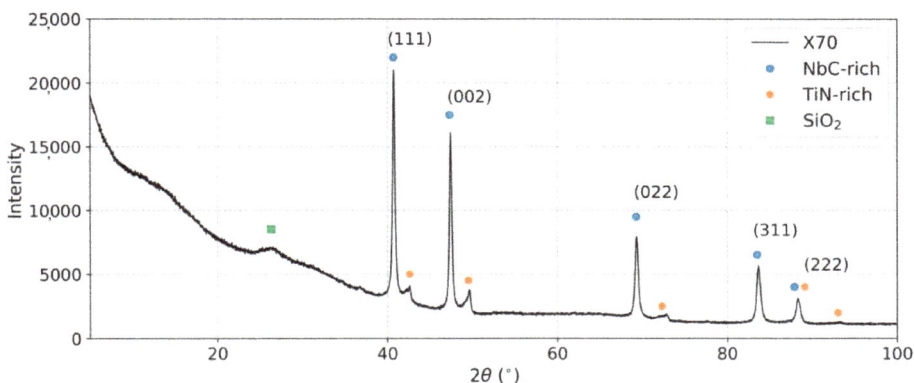

Figure 7. XRD pattern from TWX70 steel residue.

3.5. Rietveld Refinement

3.5.1. Rietveld Refinement of X70 Steel

The measured and predicted diffraction patterns for X70 are shown in Figure 8. Reasonably good agreement between the predicted and measured diffraction patterns is observed and confirms the veracity of the refinement. Included in the figure are short vertical lines which indicate the 2θ position of each peak (for each different diffracting plane) for each of the precipitate types (inset) included in the refinement. Seven types of precipitates were chosen for the fitting of the Ti/N rich diffraction peak to account for the variation in Nb and Ti atomic fraction observed in the large (TiNb)CN precipitates (Figure 4).

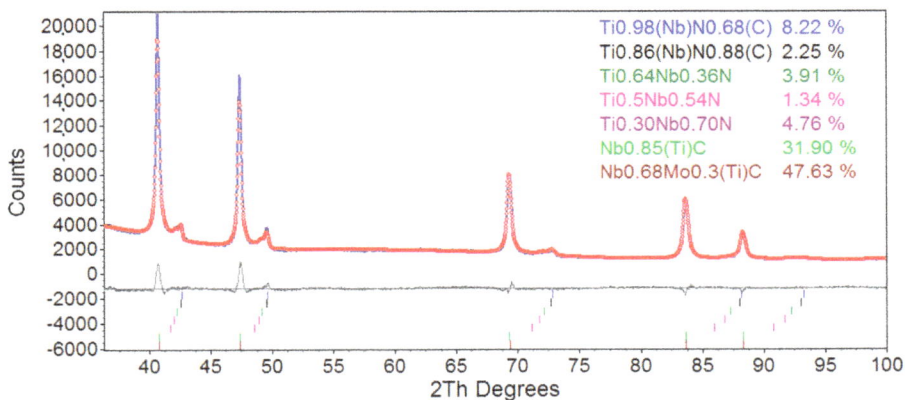

Figure 8. Comparison between measured and predicted diffraction patterns for X70 steel.

A summation of the refinement details for each precipitate type is shown in Table 4. Included in this table are the lattice parameters a (based on the 2θ peak position), the values of microstrain (ε_0) and L_{vol}, the calculated precipitate radii (based on Equation (2)), and the wt % of each precipitate type in the residue. Microstrain (or inhomogeneous strain) within the crystal lattice arises from local atomic positional distortions due to the presence of defects such as dislocations, solid solution elements, and vacancies. The microstrain levels measured for both X70 steel and TWX70 steel are relatively low. L_{vol} = 4000 nm does not represent the actual precipitate size and is the default value for the refinement program when the size of the precipitates are relatively large. The most common type of precipitate (by wt %) is nano-sized $Nb_{0.68}Mo_{0.30}Ti_{0.02}C$, followed by the slightly larger, but still nano-sized, $Nb_{0.85}Ti_{0.15}C$. These two types of precipitates account for almost 80% of the residue in the X70 steel.

Table 4. Refinement results for X70 steel.

Atomic Composition	a (Å)	ε_o (%)	L_{vol} (nm)	R (nm)	wt (%)
$Ti_{0.97}Nb_{0.03}(N_{0.68}C_{0.32})$	4.26 ± 0.02	0.14 ± 0.02	4000	-	8.2 ± 1.2
$Ti_{0.86}Nb_{0.14}(N_{0.88}C_{0.12})$	4.27 ± 0.04	0.19 ± 0.03	4000	-	2.3 ± 0.3
$Ti_{0.64}Nb_{0.36}N$	4.30 ± 0.04	0.19 ± 0.03	4000	-	3.9 ± 0.6
$Ti_{0.50}Nb_{0.50}N$	4.32 ± 0.04	0.28 ± 0.04	4000	-	1.3 ± 0.2
$Nb_{0.70}Ti_{0.30}N$	4.35 ± 0.04	0.82 ± 0.12	4000	-	4.8 ± 0.7
$Nb_{0.85}Ti_{0.15}C$	4.45 ± 0.05	0	40.3 ± 6.1	15.6 ± 2.3	31.9 ± 4.8
$Nb_{0.68}Mo_{0.30}Ti_{0.02}C$	4.45 ± 0.05	0	10.9 ± 1.6	4.2 ± 0.6	47.6 ± 7.1

3.5.2. Rietveld Refinement of TWX70 Steel

The measured and predicted diffraction patterns for TWX70 steel are shown in Figure 9. The reasonably good agreement (except for the Ti/N-rich peak at $2\theta = 49.5°$) between the predicted and measured diffraction patterns confirms the veracity of the refinement. Included in the graph are short vertical lines which indicate the 2θ position of each peak (for each different diffracting plane) for each of the precipitate types (inset) included in the refinement. Six types of precipitates were chosen for the fitting of the Ti/N rich diffraction peak to account for the variation in Nb and Ti atomic fraction observed in the large (TiNb)CN precipitates (Figure 5).

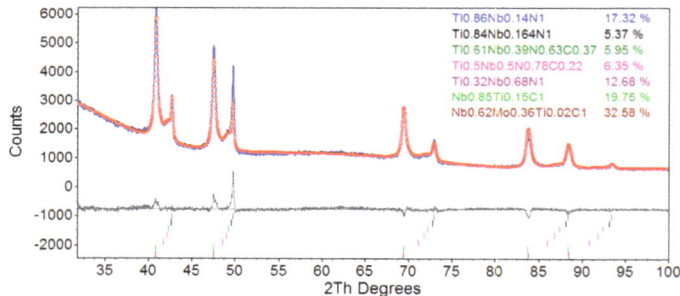

Figure 9. Comparison between measured and predicted diffraction patterns for X70 steel.

A summation of the refinement details for each precipitate type is shown in Table 5. The most common type of precipitate (by wt %) is nano-sized $Nb_{0.68}Mo_{0.30}Ti_{0.02}C$, followed by the slightly larger $Nb_{0.85}Ti_{0.15}C$ and $Ti_{0.86}Nb_{0.14}N$. Unlike the X70 steel, the former two precipitate types only account for just over 50% of the residue weight. The nitrides are much more prevalent in the TWX70 residue.

Table 5. Refinement results for TWX70 steel.

Atomic Composition	a (Å)	ε_o (%)	L_{vol} (nm)	R (nm)	wt (%)
$Ti_{0.86}Nb_{0.14}N$	4.26 ± 0.02	0.10 ± 0.02	4000	-	17.3 ± 2.6
$Ti_{0.84}Nb_{0.16}N$	4.27 ± 0.04	0.23 ± 0.03	4000	-	5.4 ± 0.8
$Ti_{0.61}Nb_{0.39}(N_{0.63}C_{0.37})$	4.29 ± 0.04	0.25 ± 0.04	4000	-	5.9 ± 0.9
$Ti_{0.50}Nb_{0.50}(N_{0.78}C_{0.22})$	4.32 ± 0.04	0.27 ± 0.04	4000	-	6.4 ± 1.0

Table 5. *Cont.*

Atomic Composition	a (Å)	ε_o (%)	L_{vol} (nm)	R (nm)	wt (%)
$Nb_{0.68}Ti_{0.32}N$	4.35 ± 0.04	0.99 ± 0.15	4000	-	12.7 ± 1.9
$Nb_{0.84}Ti_{0.16}C$	4.44 ± 0.05	0	28.9 ± 4.3	11.2 ± 1.7	19.8 ± 2.9
$Nb_{0.62}Mo_{0.36}Ti_{0.02}C$	4.44 ± 0.05	0	13.2 ± 2.0	5.1 ± 0.8	32.6 ± 4.9

3.6. Lattice Parameter Comparison

The measured lattice parameters provided by the refinement (Tables 4 and 5) are compared to the predicted lattice parameter values based on a linear interpolation (Vegard's law) between the known lattice sizes of NbC and TiC, and NbN and TiN, as a function of Ti concentration. The red line in Figures 10a and 10b represents the variation in lattice parameter of $(Ti_xNb_{(1-x)})C$ as a function of Ti content (i.e., the value of x). The blue line in each figure represents the variation in lattice parameter of $(Ti_xNb_{(1-x)})N$ as a function of Ti content (i.e., the value of x). The lattice parameters predicted by Rietveld refinement, for particular precipitate atomic compositions (represented by filled circles), agree reasonably well with the interpolated lattice parameter values. The only exception is for the smallest precipitates (i.e., predicted atomic fraction of Ti is <0.05) where Mo is present in the structure.

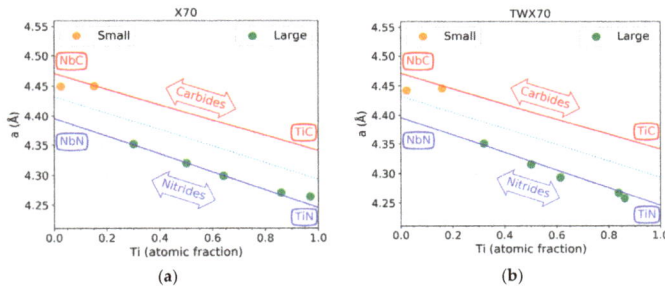

Figure 10. Lattice parameter as a function of Ti content for (**a**) X70 steel and (**b**) TWX70 steel.

4. Discussion

4.1. Mass Balance of Elements: Comparison between ICP and QXRD

A mass balance comparison, between the ICP data and the relative amount and atomic composition of each precipitate type calculated through Rietveld refinement, for Ti, Nb, and Mo was undertaken (Figure 11). The amount of each element in the precipitates from Rietveld refinement was calculated based on the size of the steel sample, the weight of the residue, the atomic composition, and the wt % determined by Rietveld refinement (Tables 4 and 5). The ICP measurements independently confirm the veracity of Rietveld refinement in determining the element distribution in the precipitates for microalloyed steels.

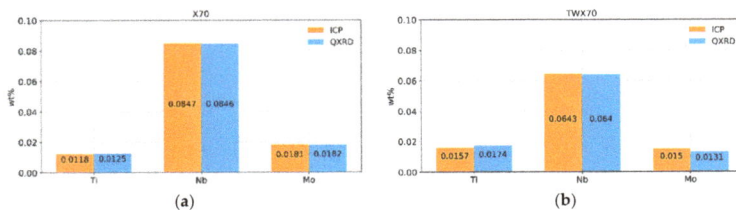

Figure 11. Comparison of Ti, Nb, and Mo amounts in precipitates for ICP and QXRD for (**a**) X70 steel and (**b**) TWX70 steel.

4.2. Nano Sized Precipitates

The sizes and atomic compositions (Nb, Mo, and Ti) of the nano-scale precipitates measured using SEM-EDX analysis are compared with Rietveld refinement values in Table 6. The mean size of the precipitates measured by SEM imaging for X70 steel was 7.95 nm, with the range of values extending from 3.6 to 15.2 nm. The distribution of measured precipitate sizes is similar to the nano precipitate sizes predicted by Rietveld refinement (4.2 nm and 15.6 nm). The mean sizes of nano-scale precipitates measured by SEM for TWX70 steel were 4.4 and 10.0 nm, with the range of values extending from 4.4 to 13.9 nm. The distribution of measured precipitate sizes correlates with the nano precipitate sizes predicted by Rietveld refinement (5.1 nm and 11.2 nm). The reasonable agreement between the SEM-measured values and those predicted by Rietveld refinement indicates that the latter technique is suitable for quantifying the mean size of nano-sized precipitates in microalloyed steels. The disadvantage of Rietveld refinement measurements is that only a mean value is provided (not a distribution). Conversely, the advantage of Rietveld refinement measurement is that a large number of precipitates are sampled in the residue versus a significantly smaller number measured directly by SEM.

The Nb composition predicted for the nano-sized precipitates compares reasonably well with the mean composition measured using EDX analysis for both the X70 and TWX70 steels (Table 6). However, the Ti composition is under-predicted, and the Mo level is over-predicted when compared with the finest (i.e., 4.2 nm and 5.1 nm) Rietveld refinement predicted precipitates. This difference may be partially attributed to the inclusion of all nano-sized precipitates (including Mo-free ones) in the determination of the mean EDX composition.

The volume percentage of each nano-sized precipitate type predicted by Rietveld refinement (V_f) and the total volume of percentage ($V_{f\text{-}Total}$) of the nano-sized precipitates in each steel are shown in Table 6. The X70 steel exhibits almost double the amount of nano-sized precipitates compared with the TWX70 steel. This difference may be partially attributed to the higher Nb content of the former.

Table 6. Summary of Rietveld refinement and SEM-EDX analysis for nano-sized precipitates.

Steel	Rietveld	EDX Composition (Mean at. Fraction)			R_{QXRD}	R_{SEM}	V_f	$V_{f\text{-}Total}$	σ_{ppt}
		Nb	Ti	Mo	(nm)	(nm)	(%)	(%)	(MPa)
X70	$Nb_{0.68}Mo_{0.30}Ti_{0.02}C$	0.79	0.16	0.05	4.2	7.95	0.066	0.113	135.1
	$Nb_{0.85}Ti_{0.15}C$				15.6	-	0.047		
TWX70	$Nb_{0.62}Mo_{0.36}Ti_{0.02}C$	0.66	0.18	0.16	5.1	4.4	0.039	0.064	133.2
	$Nb_{0.84}Ti_{0.16}C$				11.1	10.0	0.025		

4.3. Large Precipitates

Table 7 shows the volume fractions (%) for each composition of large precipitates in X70 and TWX70 steels and the total volume fraction. The total volume fraction calculated by Rietveld refinement for large precipitates in TWX70 steel is almost twice as high (0.070%) as the volume fraction for large precipitates in X70 steel (0.037%). Since large N-rich precipitates would form primarily during casting and subsequent cooling, it is postulated that the difference in the amount of large precipitates between the X70 and TWX70 steels is a result of the casting process.

Table 7. QXRD large precipitate summary for X70 and TWX70 steels.

Steel	Atomic Composition	V_f (%)	Total V_f (%)
X70	$Ti_{0.97}Nb_{0.03}N_{0.68}C_{0.32}$	0.017	0.037
	$Ti_{0.86}Nb_{0.14}N_{0.88}C_{0.12}$	0.004	
	$Ti_{0.64}Nb_{0.36}N$	0.007	
	$Ti_{0.50}Nb_{0.50}N$	0.002	
	$Ti_{0.30}Nb_{0.70}N$	0.007	

Table 7. *Cont.*

Steel	Atomic Composition	V_f (%)	Total V_f (%)
TWX70	$Ti_{0.86}Nb_{0.14}N$	0.028	0.070
	$Ti_{0.84}Nb_{0.16}N$	0.009	
	$Ti_{0.61}Nb_{0.39}N_{0.63}C_{0.37}$	0.009	
	$Ti_{0.50}Nb_{0.50}N_{0.78}C_{0.22}$	0.009	
	$Ti_{0.32}Nb_{0.68}N$	0.016	

5. Conclusions

1. N_2 injection during dissolution minimized/prevented the formation of amorphous SiO_2 during the dissolution process.

2. Rietveld refinement provided reasonably accurate nano precipitate atomic compositions and size values in the X70 microalloyed steels. These results were confirmed by direct SEM-EDX and ICP measurements.

3. Similar types of precipitates (Ti/N-rich, Nb/C-rich, and Nb/Mo/C-rich) were detected in both X70 and TWX70 steels.

4. Nano-sized precipitates with higher Mo levels and lower Nb levels were observed in the TWX70 steel. The presence of the higher Mo content nano-sized precipitates is attributed to the higher Mo content and lower Nb content in the TWX70 steel.

5. Nano-sized precipitates in the thin-walled X70 steel were present in a higher volume fraction (0.113%) than in the thick-walled TWX70 steel (0.064%). This difference is consistent with a higher Nb content in the X70 steel versus the TWX70 steel.

Author Contributions: C.C. completed this work as part of an M.Sc. Degree and was involved in all aspects of producing the work presented in this paper. J.B.W., D.G.I. and H.H. contributed to the conceptualization, analysis, and writing of this paper. The latter two authors also provided formal supervision of C.C.

Funding: This research was funded through a Natural Sciences and Engineering Research Council of Canada (NSERC) Collaborative Research and Development Grant, grant number CRDPJ 501123-16.

Acknowledgments: The authors would like to thank Laurie Collins for his input and EVRAZ N.A. Inc., TCPL, and NSERC for financial support.

Conflicts of Interest: The authors declare no conflicts of interest.

Abbreviations

The following abbreviations are used in this manuscript:

QXRD	Quantitative X-ray diffraction
ICP	Inductively coupled plasma mass spectrometry
EDX	Energy dispersive X-ray spectroscopy
SEM	Scanning electron microscopy
TEM	Transmission electron microscopy
TWX70	Thick-walled (17 mm) X70 steel

References

1. Collins, L. Processing of Nb-Containing Steels by Steckel Mill Rolling. In Proceedings of the Niobium Science and Technology: Proceedings of the International Symposium Niobium, Orlando, FL, USA, 2–5 December 2001; pp. 527–542.

2. Bai, D.; Cooke, M.; Asante, J.; Dorricott, J. Process for Making High Strength Micro-Alloy Steel. U.S. Patent US 6,682,613, 27 January 2004.

3. Lu, J. Quantitative Microstructural Characterization of Microalloyed Steels. Ph.D. Thesis, University of Alberta, Alberta, AB, Canada, 2009.

4. Liu, F.; Wang, J.; Liu, Y.; Misra, R.; Liu, C. Effects of Nb and V on microstructural evolution, precipitation behavior and tensile properties in hot-rolled mo-bearing steel. *J. Iron Steel Res. Int.* **2016**, *23*, 559–565. [CrossRef]

5. Kutz, M. *Handbook of Materials Selection*; John Wiley and Sons Inc.: New York, NY, USA, 2002; p. 44.

6. Gladman, T. *The Physical Metallurgy of Microalloyed Steels*; The Institute of Materials: London, UK, 1997.

7. Elwazri, A.M.; Varano, R.; Siciliano, F.; Bai, D.; Yue, S. Characterisation of precipitation of niobium carbide using carbon extraction replicas and thin foils by FESEM. *Mater. Sci. Technol.* **2006**, *22*, 537–541. [CrossRef]

8. Wiskel, J.; Ivey, D.G.; Henein, H. The Effects of Finish Rolling Temperature and Cooling Interrupt Conditions on Precipitation in Microalloyed Steels Using Small Angle Neutron Scattering. *Metall. Mater. Trans. A* **2008**, *39B*, 116–124. [CrossRef]

9. Staron, P.; Jamnig, B.; Leitner, H.; Ebner, R.; Clemens, H. Small-angle neutron scattering analysis of the precipitation behavior in a maraging steel. *J. Appl. Crystallogr.* **2003**, *36*, 415–419. [CrossRef]

10. Perrard, F.; Deschamps, A.; Maugis, P. Modelling the precipitation of NbC on dislocations in ferrite. *Acta Mater.* **2007**, *55*, 1255–1266. [CrossRef]

11. Wiskel, J.; Lu, J.; Omotoso, O.; Ivey, D.; Henein, H. Characterization of Precipitates in a Microalloyed Steel Using Quantitative X-ray Diffraction. *Metals* **2016**, *6*, 90. [CrossRef]

12. Lu, J.; Wiskel, J.; Omotoso, O.; Henein, H.; Ivey, D. Matrix dissolution techniques applied to extract and quantify precipitates from a microalloyed steel. *Metall. Mater. Trans. A* **2011**, *42*, 1767–1784. [CrossRef]

13. Rietveld, H. Line profiles of neutron powder-diffraction peaks for structure refinement. *Acta Crystallogr.* **1967**, *22*, 151–152. [CrossRef]

14. Rietveld, H. A profile refinement method for nuclear and magnetic structures. *J. Appl. Crystallogr.* **1969**, *2*, 65–71. [CrossRef]

15. Popa, N.; Balzar, D. An analytical approximation for a size-broadened profile given by the lognormal and gamma distributions. *J. Appl. Crystallogr.* **2002**, *35*, 338–346. [CrossRef]

16. Balzar, D.; Audebrand, N.; Daymond, M.R.; Fitch, A.; Hewat, A.; Langford, J.I.; Bail, A.L.; Louer, D.; Masson, O.; McCowan, C.N.; et al. Size–strain line-broadening analysis of the ceria round-robin sample. *J. Appl. Crystallogr.* **2004**, *37*, 911–924. [CrossRef]

17. *Standard Test Method for Acid-Insoluble Content of Copper and Iron Powders*; ASTM International: West Conshohocken, PA, USA, 2015.

18. Chatelier, C. Precipitation Analysis in Microalloyed X70 Steels and Heat Treated L80 and T95 Steels. Master's Thesis, University of Alberta, Alberta, AB, Canada, 2017.

19. Coelho, A. *Topas Academic Version 4.1*; Software for Analysis of Powder Diffraction Data; Coelho Software: Brisbane, Australia, 2007.

20. Toby, B. R factors in Rietveld analysis: How good is good enough? *Powder Diffr.* **2006**, *21*, 67–70. [CrossRef]

21. Hill, R.; Howard, C. Quantitative phase analysis from neutron powder diffraction data using the Rietveld method. *J. Appl. Crystallogr.* **1987**, *20*, 467–474. [CrossRef]

© 2018 by the authors. Licensee MDPI, Basel, Switzerland. This article is an open access article distributed under the terms and conditions of the Creative Commons Attribution (CC BY) license (http://creativecommons.org/licenses/by/4.0/).

crystals

MDPI

Article

Heavily Boron Doped Diamond Powder: Synthesis and Rietveld Refinement

Igor P. Zibrov * and Vladimir P. Filonenko

Institute for High Pressure Physics RAS, Kaluzhskoe highway, Str. 14, Troitsk, Moscow 108840, Russia; filv@hppi.troitsk.ru
* Correspondence: zibrov@hppi.troitsk.ru; Tel.: +7495-8510738; Fax: +7495-8510012

Received: 9 June 2018; Accepted: 17 July 2018; Published: 19 July 2018

Abstract: Boron-doped diamonds were synthesized by the reaction of an amorphous globular carbon powder (80%) with a powder of 1,7-di (oxymethyl)-M-carborane (20%) in a 'toroid'-type high-pressure chamber at a pressure of 8.0 GPa and temperature of 1700 °C. The structure was refined by the Rietveld method according to the X-ray powder diffraction data. It was shown that the unit cell parameters of these diamonds have two discrete quantities: around 3.570 Å for small concentrations of B (~1–1.5%) and around 3.578 Å for large concentrations of B (~2–3%). The concentration of the vacancies in the diamonds exceeds the concentration of boron atoms by 2–3 fold. This fact can play an important role in the formation of the structure and in determining the physical properties of diamonds.

Keywords: diamond; synthesis; high pressure; high temperature; X-ray powder diffraction; Rietveld refinement

1. Introduction

Diamond is a unique material that is used in various industrial areas, especially in electronics [1,2]. There is active investigation into the possibility of introducing different elements into the diamond lattice. This aims to complement the outstanding properties of diamonds, including high hardness, high thermal conductivity, and high chemical resistance to acids and alkalis, with new optical and electrical characteristics. Most natural and synthetic diamonds contain nitrogen, which makes them yellow. Nitrogen acts as a substitute for carbon in diamond, with a maximum concentration of about one percent (~10^{20} cm^{-3}). The content of larger atoms (Si, Ti, Ni, Ge, and P) in the diamond lattice is less than 0.01%. They form dopant–vacancy complexes with different luminescent characteristics [3,4]. As a rule, substituting atoms slightly increases the diamond unit cell parameter.

Boron is a neighbor of carbon in the periodic table although it is rarely found in natural diamonds. Several methods have been developed for producing boron-doped diamonds (BDD) with low electrical resistance, high mobility of charge carriers, and p-type conductivity. Large single crystals are synthesized at pressures of about 5 GPa [5]. In this case, the boron is in the molten metal through which the graphite recrystallization takes place. As the concentration of boron increases, the color of diamond changes from blue to black. Various methods of growing diamond films have also made it possible to produce single crystals with a boron content of >10^{20} cm^{-3} in the lattice [6]. The electrical conductivity of BDD is most often used for the manufacture of electrodes and sensors. Both CVD films and microcrystals are used for this purpose [7–9].

With an increase in the boron concentration, the conductivity of the diamond increases. If the content of boron is more than 1 wt %, the transition of the material to a superconductor can be observed at helium temperatures. The superconductivity of BDD was first discovered in 2004 with a polycrystalline sample synthesized from a mixture of graphite and boron carbide [10].

It is known that the unit cell parameter of the diamond increases with the substitution of carbon by boron. The maximum increase was recorded in [11,12]. As a result of the graphite–diamond phase transition, polycrystals with the unit cell parameters of 3.574–3.577 Å were formed, with the content of boron in these polycrystals being estimated at 3–4 wt %.

Despite the large number of results relating to the production and study of BDD, discussions about the concentration of boron and its role in superconductivity is ongoing. This is partly due to the fact that good-quality single crystals with a high content of boron have not yet been obtained.

The goal of this work was to synthesize microcrystals of diamonds with high boron content, with subsequent refinement of their structure.

2. Materials and Methods

The starting materials for the work were amorphous globular carbon powder with particle sizes of about 25 nm (80%) and powder of 1,7-di (oxymethyl)-M-carborane with the formula HO-H_2C-$CB_{10}H_{10}$C-CH_2-OH (20%) (Aviabor, Dzerzhinsk, Russia). M-carborane decomposes at a high temperature to form active atomic boron. The powders were stirred in alcohol with application of ultrasound. The obtained mixture was pressed into pellets with the following dimensions: 3 mm in height and 5 mm in diameter. The thermobaric treatment of pellets was conducted in "toroid"-type high-pressure chambers. The scheme of the high-pressure chamber and the method were previously described in references [13,14]. The temperature in the chamber was calibrated using a W:W-Re thermocouple, with the accuracy of measurements estimated at ±25 °C. The material was treated under the following conditions: P = 8.0 GPa, T = 1700 °C for 20 s (sample N1) and 30 s (sample N2).

Crystal morphology was studied using microscopes SEM JSM-6390 JEOL (JEOL Ltd., Tokyo, Japan) and TEM JEM-2100 JEOL (JEOL Ltd., Tokyo, Japan) with accelerating voltage of 200 kV.

Micro-Raman measurements were obtained at room temperature using the TriVista 555 triple grating spectrometer with a liquid-nitrogen-cooled CCD detector. The 488 nm line of the Ar^+ ion laser was used for excitation. To avoid overheating or burning out of the samples, the laser power was kept at a minimum (approximately 0.5 mW) and the 50× objective (NA = 0.5) of Olympus BX51 microscope (Olympus, Tokyo, Japan) was used for laser focusing and the scattered light collection.

A Huber Imaging Plate Guinier camera G670 (Cu $K\alpha_1$ radiation, Huber Technology, Tutzing, Germany) was used for the phase analysis and data collection.

3. Results

The chemical interaction of amorphous globular nano-carbon with active boron leads to BDD crystallization with particle sizes of up to 5 μm in a few seconds. Phase analysis showed that both samples consist of three phases: two diamond D1 and D2 with different unit cell parameters and carbide B_4C. In this case, the diamond phases had unit cell parameters greater than that of a pure diamond (a = 3.567 Å, ICSD 76766), which is one of the indicators of a heavily BDD. This is well illustrated by the diffraction patterns in the region of the diamond peak (311) (Figure 1). Raman spectra of the obtained diamonds (Figure 2) also showed features that were characteristic of crystals with a high content of boron in the lattice. The characteristics and description of these features can be found in a previous study [15]. The difference in the spectra of Samples 1 and 2 is due to the different content of D1 and D2 (Tables 1 and 2). M-carborane contains 52.92% of boron and thus, the content of boron is 10.59% in the samples.

Diamond single crystals were formed during the thermobaric treatment of mixtures of nano-carbon with M-carborane, which have dimensions in the range of submicrons to several micrometers. The morphology of the crystals is shown in Figure 3.

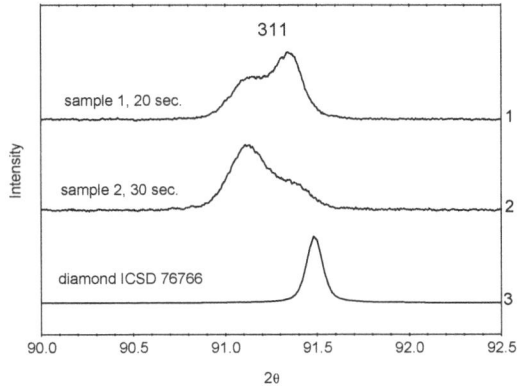

Figure 1. Diffraction patterns in the region of the diamond peak (311): 1—sample N1, 2—sample N2, 3—diamond ICSD 76766.

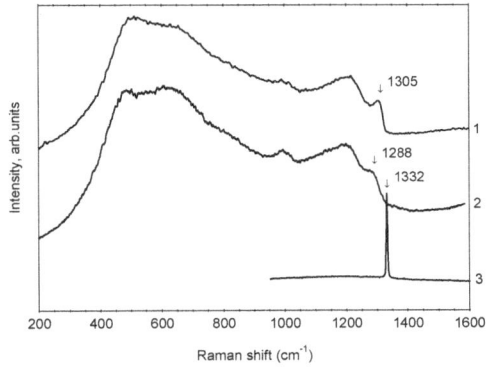

Figure 2. Raman spectra: 1—Sample N1; 2—Sample N2; 3—ordinary microdiamond.

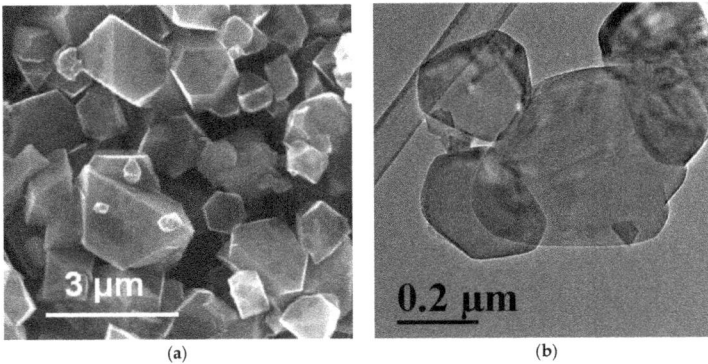

Figure 3. Morphology of boron doped diamond crystals. Synthesis from a mixture of globular nano-carbon and M-carborane. (**a**) SEM; (**b**) TEM.

Rietveld full-profile refinement was performed with the program GSAS [16,17]. The observed diffractometer data and the difference between the observed and calculated data are shown in Figures 4 and 5. The crystal data and structure refinement are shown in Tables 1 and 2. The unit cell parameters and the weight fraction in the mixture were refined for B_4C, while the atomic parameters were taken from [18].

We can calculate the content of boron in diamonds as we know the weight fraction of boron carbide in the sample (Tables 1 and 2). Boron carbide has a broad homogeneity region. When changing the composition of carbide from B_8C to B_4C, the parameters of the unit cell are significantly reduced. If we compare the unit cell volumes of boron carbide in Samples 1 and 2 (Tables 1 and 2) with the results obtained in [19], it can be concluded that the composition is B_4C. The boron concentration in the diamonds were calculated to be 1.93% and 1.51% for Samples 1 and 2, respectively. Obviously, the calculations are carried out under the assumption that all boron from M-carborane passes into diamonds and carbide.

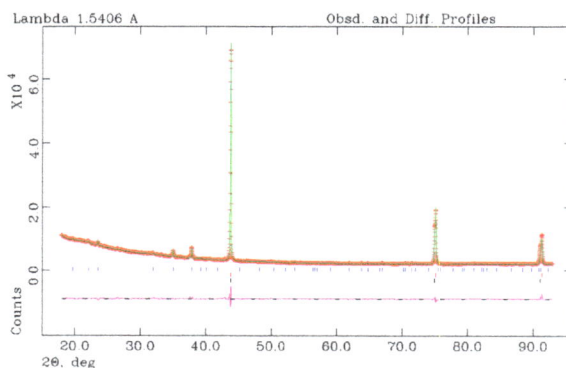

Figure 4. X-ray Rietveld refinement of the sample N1. The observed (+), calculated (solid line) and difference between observed and calculated (bottom curve) powder diffraction profiles. The positions of all allowed Bragg reflections are indicated by the rows of vertical tick marks: B_4C upper and diamonds lower marks.

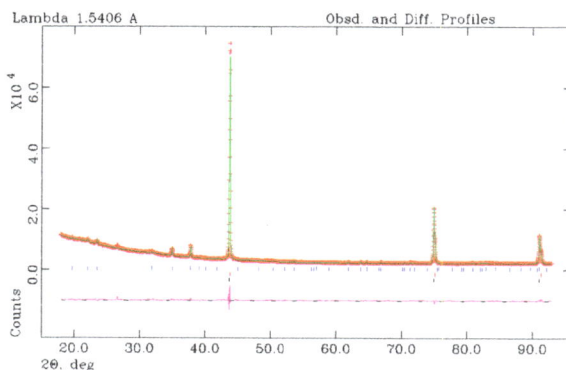

Figure 5. X-ray Rietveld refinement of the sample N2. The observed (+), calculated (solid line) and difference between observed and calculated (bottom curve) powder diffraction profiles. The positions of all allowed Bragg reflections are indicated by the rows of vertical tick marks: B_4C upper and diamonds lower marks.

Table 1. Experimental details for D1, D2, and B_4C (sample N1).

Chemical Formula	C (D1)	C (D2)	B_4C
Chemical formula weight	12.01	12.01	164.74
Space group	$F\,d-3\,m$	$F\,d-3\,m$	$R-3\,m$
a (Å)	3.57831(1)	3.57141(1)	5.5862(1)
b (Å)			5.5862(1)
c (Å)			12.0484(4)
γ (°)			120.00
V (Å^3)	45.818(0)	45.553(0)	325.611(13)
Z	8	8	3
D_x (Mg m^{-3})	3.266	3.388	2.506
Weight fraction, %	34.44(10)	54.51(5)	11.05(7)
Radiation type		Cu Kα_1	
Wavelength (Å)		1.5405981	
Temperature (K)		293	
Data collection			
Diffractometer	Imaging Plate Guinier camera G670, Huber		
Refinement		GSAS	
R_F	0.0091	0.0164	
R_P		0.0184	
R_{WP}		0.0276	

Table 2. Experimental details for D1, D2, and B_4C (sample N2).

Chemical Formula	C (D1)	C (D2)	B_4C
Chemical formula weight	12.01	12.01	164.74
Space group	$F\,d-3\,m$	$F\,d-3\,m$	$R-3\,m$
a (Å)	3.57708(1)	3.57010(2)	5.58783(9)
b (Å)			5.58783(9)
c (Å)			12.0477(3)
γ (°)			120.00
V (Å^3)	45.770(0)	45.503(1)	325.78(1)
Z	8	8	3
D_x (Mg m^{-3})	3.382	3.488	2.505
Weight fraction, %	66.51(4)	21.90(11)	11.59(7)
Radiation type		Cu Kα_1	
Wavelength (Å)		1.5405981	
Temperature (K)		293	
Data collection			
Diffractometer	Imaging Plate Guinier camera G670, Huber		
Refinement		GSAS	
R_F	0.0098	0.0138	
R_P		0.0175	
R_{WP}		0.0250	

The conducted refinement of the occupancy of carbon positions by boron in the Sample 1 gave a boron concentration of 29.8(8)% in D1 and 13.5(8)% in D2, which is higher than the calculated values. Similar values were obtained for Sample 2. This result clearly shows the deficiency of an electron density in the carbon positions. It can only be explained by the presence of vacancies since boron reduces the electron density by only one electron compared to the density on carbon, and a vacancy reduces it by six electrons. Table 3 shows the results of the refinement under the assumption that both boron atoms and vacancies are present in diamonds. In the refinement, the boron concentration was fixed: 3.12% for D1 and 1.56% for D2, which corresponded to the final total of 1.93% (it was assumed that the concentration of boron in D1 is two times higher than that in D2).

Table 3. Fractional atomic coordinates, site occupancy, and isotropic displacement parameters U_{iso} (Å^2) for D1 and D2 (Sample 1).

Phase	ATOM	Site	OCC	x	y	z	U_{iso}
D1	C	(8a)	0.910(3)	0.125	0.125	0.125	0.0166(1)
	B	(8a)	0.0312	0.125	0.125	0.125	0.0166(1)
D2	C	(8a)	0.953(3)	0.125	0.125	0.125	0.0141(1)
	B	(8a)	0.0156	0.125	0.125	0.125	0.0141(1)

4. Discussion

Table 3 shows that the concentration of vacancies in D1 is approximately 6%, while this is 3% in D2. This essentially means that each boron atom is associated with two vacancies in both diamonds. The same refinement for Sample 2 gives around three vacancies per B atom in both diamonds. It should be noted that we have prepared about 10 samples under different P–T conditions and different starting mixtures. The unit cell parameters of the obtained diamonds were either about 3.570 Å or 3.578 Å. Two different concentrations of boron can be associated with the presence of stable electronic states: semiconducting and metallic. In this case, the transition between states can be considered as an electronic Mott transition.

The refinement of the structure does not provide the reason for the discretization of the unit cell parameter of the diamonds although this shows the significant influence of vacancies as their concentration is 2–3 times higher than the boron concentration.

5. Conclusions

For the first time, by using Rietveld full-profile refinement, it is shown that the unit cell parameter of a diamond with boron has two discrete values: around 3.570 Å for small concentrations of B (~1–1.5%) or around 3.578 Å for large concentrations of B (~2–3%). It is also shown that these processes can be influenced by vacancies, the concentration of which is 2–3 times higher than the concentration of boron in diamonds.

Author Contributions: V.P.F. finished the HP-HT synthesis, I.P.Z. performed X-ray study. Both authors analyzed and discussed the results.

Funding: This work was financially supported by RFBR grant 17-02-01285 a.

Acknowledgments: The authors thank Lyapin S.G. for the Raman spectra measurements and Trenikhin M.V. for the electron microscopy studies.

Conflicts of Interest: The authors declare no conflicts of interest.

References

1. Scott, D.E. The history and impact of synthetic diamond cutters and diamond enhanced inserts on the oil and gas industry. *Ind. Diamond Rev.* **2006**, *1*, 48–58.
2. Vavilov, V.S. Diamond in solid state electronics. *Phys. Uspekhi* **1997**, *1*, 15–20. [CrossRef]
3. Nadolinny, V.; Komarovskikh, A.; Palyanov, Y. Incorporation of large impurity atoms into the diamond crystal lattice: EPR of split-vacancy defects in diamond. *Crystals* **2017**, *7*, 237. [CrossRef]
4. Ekimov, E.A.; Lyapin, S.G.; Boldyrev, K.N.; Kondrin, M.V.; Khmelnitskiy, R.; Gavva, V.A.; Kotereva, T.V.; Popova, M.N. Germanium–vacancy color center in isotopically enriched diamonds synthesized at high pressures. *JETP Lett.* **2015**, *102*, 701–706. [CrossRef]
5. Blank, V.D.; Kuznetsov, M.S.; Nosukhin, S.A.; Terentiev, S.A.; Denisov, V.N. The influence of crystallization temperature and boron concentration in growth environment on its distribution in growth sectors of type IIb diamond. *Diam. Relat. Mater.* **2007**, *16*, 800–804. [CrossRef]

6. Issaoui, R.; Achard, J.; Silva, F.; Tallaire, A.; Tardieu, A.; Gicquel, A.; Pinault, M.A.; Jomard, F. Growth of thick heavily boron-doped diamond single crystals: Effect of microwave power density. *Appl. Phys. Lett.* **2010**, *97*, 182101. [CrossRef]

7. Bautze, T.; Mandal, S.; Williams, O.A.; Rodiere, P.; Meunier, T.; Bauerle, C. Superconducting nano-mechanical diamond resonators. *Carbon* **2014**, *72*, 100–105. [CrossRef]

8. Tago, S.; Ochiai, T.; Suzuki, S.; Hayashi, M.; Kondo, T.; Fujishima, A. Flexible boron-doped diamond (bdd) electrodes for plant monitoring. *Sensors* **2017**, *17*, 1638. [CrossRef] [PubMed]

9. Ochiai, T.; Tago, S.; Hayashi, M.; Hirota, K.; Kondo, T.; Satomura, K.; Fujishima, A. Boron-doped diamond powder (BDDP)-based polymer composites for dental treatment using flexible pinpoint electrolysis unit. *Electrochem. Commun.* **2016**, *68*, 49–53. [CrossRef]

10. Ekimov, E.A.; Sidorov, V.A.; Bauer, E.D.; Melnik, N.N.; Curro, N.J.; Thompson, J.D.; Stishov, S.M. Superconductivity in diamond. *Nature* **2004**, *428*, 542–545. [CrossRef] [PubMed]

11. Dubrovinskaia, N.; Eska, G.; Sheshin, G.A.; Braun, H. Superconductivity in polycrystalline boron-doped diamond synthesized at 20 GPa and 2700 K. *J. Appl. Phys.* **2006**, *99*, 033903. [CrossRef]

12. Ekimov, E.A.; Ralchenko, V.; Popovich, A. Synthesis of superconducting boron-doped diamond compacts with high elastic moduli and thermal stability. *Diamond Relat. Mat.* **2014**, *50*, 15–19. [CrossRef]

13. Filonenko, V.P.; Zibrov, I.P. High-pressure phase transitions of M_2O_5 (M=V, Nb, Ta) and thermal stability of new polymorphs. *Inorg. Mater.* **2001**, *37*, 953–959. [CrossRef]

14. Zibrov, I.P.; Filonenko, V.P.; Werner, P.-E.; Marinder, B.-O.; Sundberg, M. A new high-pressure modification of Nb_2O_5. *J. Solid State Chem.* **1998**, *141*, 205–211. [CrossRef]

15. Szirmai, P.; Pichler1, T.; Williams, O.A.; Mandal, S.; Bäuerle, C.; Simon, F. A detailed analysis of the Raman spectra in superconducting boron doped nanocrystalline diamond. *Phys. Status Solidi B* **2012**, *249*, 2656–2659. [CrossRef]

16. Larson, A.C.; Von Dreele, R.B. *General Structure Analysis System (GSAS)*; Report LA-UR-86-748; Los Alamos National Laboratory: Los Alamos, NM, USA, 1987.

17. Toby, B.H. EXPGUI, a graphical user interface for GSAS. *J. Appl. Crystallogr.* **2001**, *34*, 210–213. [CrossRef]

18. Kwei, G.H.; Morosin, B. Structures of the boron-rich boron carbides from neutron powder diffraction: Implications for the nature of the inter-icosahedral chains. *J. Phys. Chem.* **1996**, *100*, 8031–8039. [CrossRef]

19. Ponomarev, V.I.; Kovalev, I.D.; Konovalikhin, S.V.; Vershinnikov, V.I. Ordering of Carbon atoms in boron carbide structure. *Cryst. Rep.* **2013**, *58*, 422–427. [CrossRef]

© 2018 by the authors. Licensee MDPI, Basel, Switzerland. This article is an open access article distributed under the terms and conditions of the Creative Commons Attribution (CC BY) license (http://creativecommons.org/licenses/by/4.0/).

crystals

MDPI

Article

Does Thermosalient Effect Have to Concur with a Polymorphic Phase Transition? The Case of Methscopolamine Bromide

Teodoro Klaser [1], Jasminka Popović [2], José A. Fernandes [1], Serena C. Tarantino [3], Michele Zema [3] and Željko Skoko [1,*]

[1] Department of Physics, Faculty of Science, University of Zagreb, 10000 Zagreb, Croatia; tklaser@phy.hr (T.K.); jafernandes1974@gmail.com (J.A.F.)
[2] Ruđer Bošković Institute, Bijenička 54, 10000 Zagreb, Croatia; jpopovic@irb.hr
[3] Department of Earth and Environmental Sciences, University of Pavia, 27100 Pavia, Italy; serenachiara.tarantino@unipv.it (S.C.T.); michele.zema@unipv.it (M.Z.)
* Correspondence: zskoko@phy.hr; Tel.: +385-1-4605-813

Received: 14 June 2018; Accepted: 19 July 2018; Published: 21 July 2018

Abstract: In this paper, we report for the first time an observed thermosalient effect that is not accompanied with a phase transition. Our experiments found that methscolopamine bromide—a compound chemically very similar to another thermosalient material, oxitropium bromide—exhibited crystal jumps during heating in the temperature range of 323–340 K. The same behavior was observed during cooling at a slightly lower temperature range of 313–303 K. Unlike other thermosalient solids reported so far, no phase transition was observed in this system. However, similar to other thermosalient materials, methscolopamine showed unusually large and anisotropic thermal expansion coefficients. This indicates that the thermosalient effect in this compound is caused by a different mechanism compared to all other reported materials, where it is governed by sharp and rapid phase transition. By contrast, thermosalient effect seems to be a continuous process in methscolopamine bromide.

Keywords: thermosalient materials; jumping crystals; scopolamine bromide; negative thermal expansion; HT-XRPD

1. Introduction

Materials that exhibit mechanical response to external stimuli (heat or light) in the form of jumping, bursting, curling, bending, etc. are at the frontier of research into potential actuators at the nanoscale. Molecular crystals that exhibit such behavior are extremely interesting from not only a scientific aspect, but also from a technological point of view due to the rapidness of their actuation. The rapidness, controllability, and high efficiency rate of energy transduction make such materials excellent candidates for production of smart medical devices or implants, artificial muscles, biomimetic kinetic devices, electromechanical devices, actuators, materials for electronics, and heat sensitive sensors [1,2]. Among mechanically responsive single crystals, thermosalient (TS) materials are a class that stand out in particular [3–11]. Thermosalient materials, colloquially known as "jumping crystals", are materials that exhibit mechanical motion during heating/cooling, thus transforming thermal energy into mechanical work. Although thermosalient compounds are known to belong to different classes of materials—from simple organic molecules to organometallic compounds, from metal complexes all the way to inorganic solid—all thermosalient compounds exhibit three common features: 1) crystallinity, 2) negative thermal expansion for at least of one of the cell parameters, and 3) a phase transition concomitant with a sudden change of cell parameters. Even though the liberation of crystal stress during the phase transition is the most probable explanation for this, a full elucidation that would be valid for all thermosalient systems is still not established. Recent studies have shown that negative compressibility might be the driving force for thermosalient effect [12]. Additionally, our own theoretical calculations—performed for the first time on thermosalient materials—showed that thermosalient effect is caused by the softening of the low-energy phonon [12]. Despite the fact that the mechanism beyond the thermosalient phenomena is not yet completely resolved, practical applications are slowly emerging. For example, the movement or breakage of a crystal of 1,2,4,5-tetrabromobenzene coated with silver has been used for the preparation of a fuse that is activated by increasing temperature [13].

In the light of all the new cognitions on thermosalient behavior and several years after we first published a comprehensive study on TS materials, we again return to the anticholinergic agent oxitropium bromide, which sparked our initial interest in this field [14]. This compound has two polymorphs (A and B), both belonging to the space group $P2_12_12_1$, with subtle differences in the molecular conformations. The thermally induced single-crystal-to-single-crystal polymorphic transition from phase A to phase B occurs at T = 331 K. Phase transition is characterized by anisotropic changes in the cell parameters (Δa: +1%; Δb: +11%; Δc: −7%; ΔV: +4%) accompanied by jumping of crystals up to 2 cm in height while maintaining the crystal integrity. Considering all the accumulated knowledge about TS materials, this paper endeavors to answer a new question: What would happen if we introduce a small, subtle change in the chemical composition of oxitropium bromide? We do this by examining methscopolamine bromide, a compound chemically very similar to oxitropium bromide. The only difference between oxitropium and scopolamine bromide is the aza-tricyclic part of the molecules; in oxitropium bromide, the quaterny nitrogen atom has both methyl and ethyl group as substituents, whereas in methscopolamine bromide, the nitrogen atom bears two methyl groups (Figure 1).

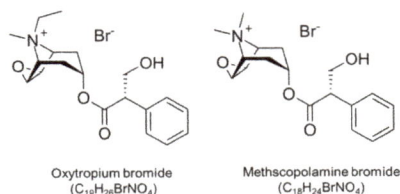

Oxytropium bromide
($C_{19}H_{26}BrNO_4$)

Methscopolamine bromide
($C_{18}H_{24}BrNO_4$)

Figure 1. Chemical structures of oxitropium bromide and methscopolamine bromide.

The crystal structure of scopolamine bromide has been thoroughly described by Glaser et al. (orthorhombic space group $P2_12_12_1$ at 293(2) K: a = 7.0403(8), b = 10.926(2), c = 23.364(5) Å, V = 1797.2(6) Å^3, Z = 4) [15]. In the present paper, we use variation temperature study to show that methscopolamine bromide exhibits thermosalient behavior without any associated phase transition. This is contrary to all previously reported thermosalient materials. Indeed, methscopolamine bromide is the only known compound in which TS behavior is not concomitant with a phase transition. Results presented in this work show that thermosalient effect is possible even without phase transition and that despite all the knowledge we have already acquired in this field, there is still a long way before the mystery of thermosalient effect is resolved.

2. Materials and Methods

Methscopolamine bromide used for the experiments was purchased from Sigma Aldrich (Sigma Aldrich, Steinheim, Germany) (>99%, HPLC). It was used as received.

2.1. X-Ray Powder Diffraction (XRPD)

Temperature-induced structural changes were tracked by in situ HT variable temperature (VT) XRPD using a Philips PW 1710 diffractometer (Philips, Almelo, The Netherlands) equipped with high temperature chamber. Diffraction patterns were collected in 2θ range 5–50° using monochromatized CuKα radiation (monochromator: graphite). Data were collected in temperature range of 300 K–458 K. Crystal structures were refined by the Rietveld method using HighScore Xpert Plus (Version 4.5, March 2016). Thermal expansion coefficients were calculated from the refined unit cell parameters obtained from variable temperature diffraction data. Linear axial thermal expansion coefficients along the principal axes were calculated using the PASCal software [16].

2.2. Thermal Analysis

Differential Scanning Calorimetry (DSC) was carried out on Mettler Toledo DSC 822e instrument (Mettler Toledo, Columbus, OH, USA) in dynamic helium atmosphere (flow rate 50 mL/min) on the pristine samples in the temperature range between 298 K and 573 K.

2.3. Hot-Stage Microscopy

Mechanical behavior during heating/cooling was recorded using Nikon Eclipse LV150NL (Nikon, Tokyo, Japan) optical microscope equipped with a Linkam THMS600 hot-stage and OPTOCAM-II color camera with a resolution of 1600 × 1200 pixels. Crystal behavior was monitored in the temperature interval from room temperature to the melting point (503 K).

3. Results and Discussion

3.1. Hot-Stage Microscopy and Thermal Analysis

Hot-stage experiments were conducted on methscopolamine bromide crystals in air. During the first heating run, crystals started jumping at ~323 K. Jumping of crystals continued up to ~340 K. Not all the crystals jumped in the course of heating, with approximately 50% doing so. Interestingly, during the cooling run, crystals jumped again, starting at the temperature of ~313 K and finishing at around 303 K. During the second heating run, several crystals jumped again at the same temperature, but the number of crystals jumping was much smaller compared to the first heating. Hot-stage experiments were performed many times with different parameters. This included heating/cooling rate (10 to 50 K/min), number of cooling runs (up to five), crystals monitored for jumping when heating/cooling was stopped at selected temperatures (323 K, 333 K, and 343 K), number of crystals, different crystal sizes, and crystal orientation.

Several features were observed:

- Jumping (in terms of number of crystal that jumped, their frequency, or strength of the jumps) did not depend on the heating/cooling rate.
- Jumping did not depend on the size, shape, or orientation of the crystals. As expected, jumps of the smaller crystals were more forceful, whereas the more massive crystals would only slightly move or turn over to another facet.
- The number of crystals that jumped decreased drastically with consecutive heating/cooling runs. For example, if 10 crystals jumped during the first heating run, only 2–3 would jump in the second heating run.
- Crystals continued to jump sporadically when temperature was maintained for some time within the jumping temperature interval between 323 K and 333 K. Time period of jumping depended on the temperature. At 323 K, crystals continued to jump for 10 minutes, whereas the jumps ceased after 1–2 minutes at 343 K.
- No breaking or cracking of the crystals were observed during the jumping.
- Overall, the jumps of scopolamine bromide crystals were less energetic compared to the crystals of oxitropium bromide.

Figure 2 shows crystals of scopolamine bromide before jumping (left panel, taken during heating at the temperature of 315 K) and after jumping (right panel, taken during heating at the temperature of 345 K). Crystals that exhibited mechanical motion are marked with blue, green, orange, red, and purple rectangles. Blue, orange and red rectangles mark crystals that jumped off the hot-stage and left the recorded area. The green rectangle marks a large crystal that flipped to the other side but remained in more or less the same position. The purple rectangle marks a crystal that rotated around its axis but was held at the same place by the larger crystal on top of it.

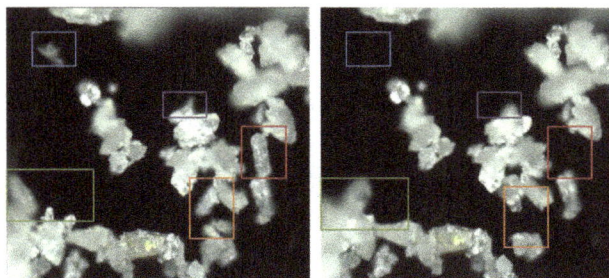

Figure 2. Crystals of scopolamine bromide before jumping (**left panel**) and after jumping (**right panel**).

Videos of crystals of methscopolamine bromide jumping during heating and cooling are provided in the supporting information (Video S1 and Video S2). DSC measurements were performed on the pristine crystals in order to reveal the phase transitions of methscopolamine bromide. In addition, thermal behavior of methscopolamine bromide was examined in the temperature interval between 298 K and 573 K. Much to our surprise, as can be seen from Figure S1, no maxima corresponding to phase transitions were observed before the melting point (around 503 K).

3.2. In Situ Variable Temperature X-ray Powder Diffraction (VT XRPD)

In situ XRPD measurements on methscopolamine bromide were performed in the temperature range of 300 K to 458 K, as shown in Figure 3.

Careful examination of powder diffraction data collected as a function of temperature (prior to any calculations) revealed quite pronounced shift in diffraction lines. Depending on the *hkl* index, diffraction lines shifted towards a lower or a higher 2θ angle; this was the first indication that scopolamine bromide is characterized by anisotropic thermal expansion. This is best illustrated in the narrow 2θ range between 16° and 17° shown in Figure 3; the diffraction line 020 shifted to higher values of 2θ angle with the increase in temperature, while the diffraction line 112 shifted towards lower angles, indicating that the unit cell of methscopolamine bromide decreased in *b*-direction and expanded in *c*-direction during heating. The shift of the 112 peak was greater compared to the 002, suggesting that the absolute value of the thermal expansion coefficient along the *c* axis was larger than along the *b* axis.

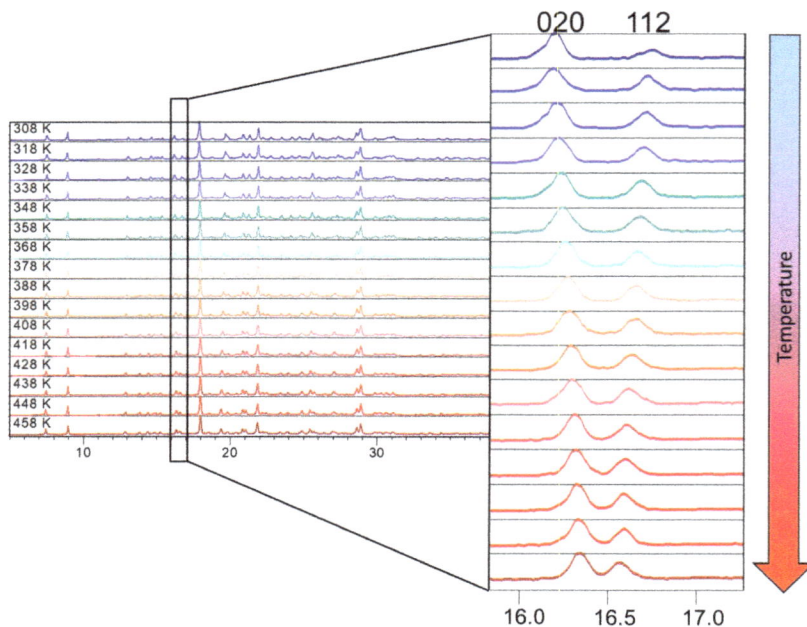

Figure 3. In situ variable temperature X-ray diffraction patterns of methscopolamine bromide in the temperature range of 308 K to 458 K. The 2θ range from 16° to 17° is enlarged and clearly shows the thermally induced shifts of the diffraction lines 020 and 112.

In order to get an insight into the temperature-induced structural changes of scopolamine bromide, Rietveld refinements were carried out on datasets collected in the temperature range of 308–458 K. Refinements were carried out starting from the structural model of Glaser [15]. Figure 4 shows the Rietveld refinements of scopolamine bromide at 308 K and 458 K; refined crystal packings at 308 K and 458 K are showed in overlap manner. Figure 4c,d shows that besides the anisotropic expansion of cell, there were no significant changes in the crystal packing. An in situ single-crystal X-ray diffraction study is also underway to get a better understanding of the molecular motion of SMB with changing temperature. Preliminary data seems to confirm the results of the Rietveld refinements. Contrary to methscopolamine bromide, quite pronounced differences in molecular structures and crystal packings have been noted in the case of oxitropium bromide between low-temperature phase A and high-temperature phase B. In fact, the proposed mechanisms beyond the thermosalient effect—not only for oxitropium bromide but for most thermosalient materials—was based on the following scenario: Heating of low-temperature polymorph is accompanied by various conformational changes of the molecule itself and consequently causes continuous changes of the packing. The shear strain caused by the distortion of the unit cell, which is almost always very anisotropic, is accrued to the point where it overweighs the cohesive interactions. At this point, accumulated stress is released and low-temperature polymorph abruptly switches to high-temperature phase. However, this premise gives rise to a new question about the rationale behind the thermosalient effect in scopolamine bromide, a material that exhibits no conformational changes during heating. A somewhat similar situation has been noted by researchers for 1,2,4,5-tetrabromobenzene. This compound does have a phase transition, but the difference in the crystal structures and intermolecular interaction energies of the low- and high-temperature phases is too small to be able to account for the large stress that arises over the course of the transformation [17].

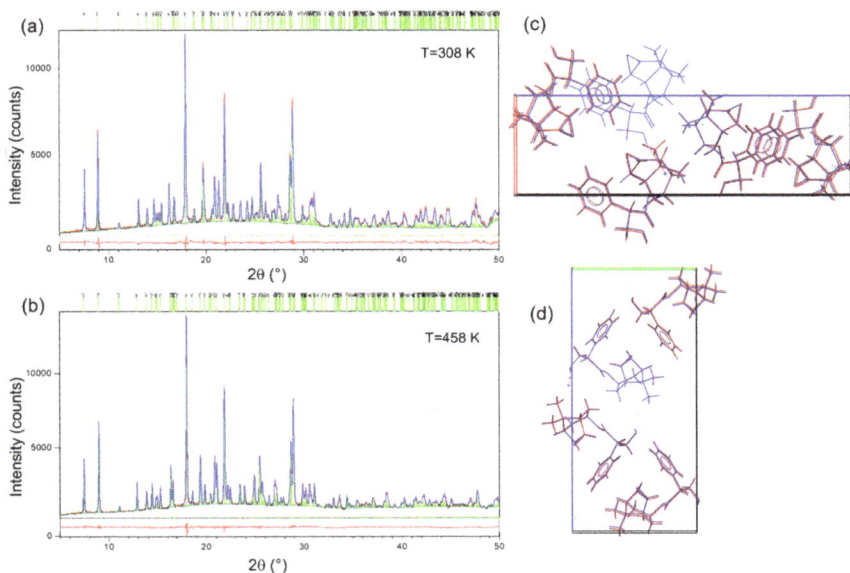

Figure 4. (**a**) Rietveld refinement fit for data collected at 308 K. (**b**) Rietveld refinement fit for data collected at 458 K. Experimental data are given as red line, calculated diffraction pattern as blue line, and the difference curves are given in red underneath the patterns. The green vertical lines represent positions of Bragg reflections of scopolamine bromide. (**c**) Overlap of crystal packings of scopolamine bromide at 308 K (blue) and 458 K (red) viewed along *b*-direction. (**d**) Overlap of crystal packings of scopolamine bromide at 308 K (blue) and 458 K (red) viewed along *a*-direction.

In the course of Rietveld refinement, unit cell parameters were determined and refined. Thermally induced changes of the unit cell parameters of methscopolamine bromide are shown in Figure 5 together with thermal expansivity indicatrix. As evident from the Figure 5, cell parameters *a* and *c* showed positive and linear thermal expansion in the investigated temperature range, while negative expansion occurred along *b*. A linear model was used to calculate the axial thermal expansion coefficients, although a slight deviation from linearity was observed along *b*. Negative thermal expansion was observed in several inorganic compounds, but it was very rare for organic compounds [18]. Also, the values of the thermal expansion coefficients were larger than is usual for molecular solids. Typical values for molecular solids are in the range of 0–20 $\times 10^{-6}$ K^{-1} [19], and our values were six times higher for the positive expansion than the maximum typical values (along *a* and *c* axis) and two times higher for the negative expansion (along *b* axis).

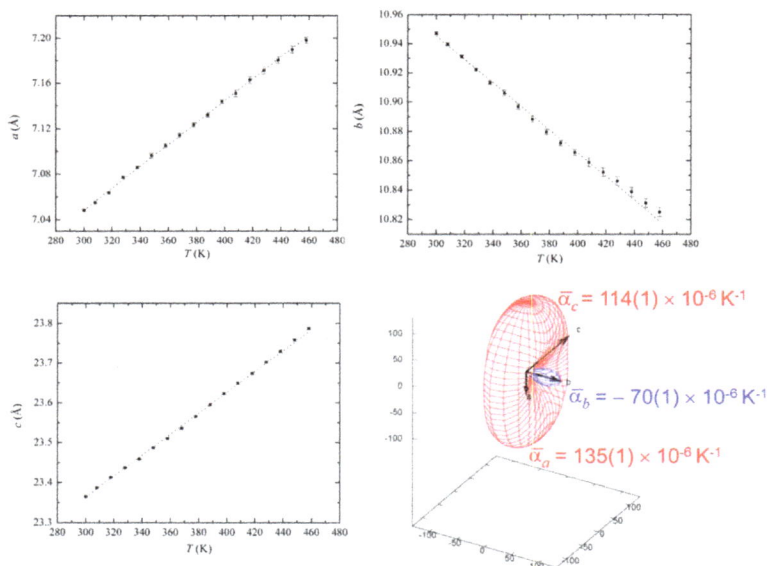

Figure 5. Temperature dependence of the unit cell parameters *a*, *b* and *c* of methscopolamine bromide. Thermal expansivity indicatrix is showing large anisotropic thermal expansion.

Since phase transition can be ruled out as a governing force for thermosalient effect in scopolamine bromide, the only plausible explanation can be that it is caused by the extremely large and anisotropic thermal expansion. Large anisotropic thermal expansion is characteristic of all reported thermosalient materials, and this extraordinary feature is shared with methscopolamine bromide as well. At the same time, this is also what differentiates them from non-thermosalient materials. All thermosalient materials exhibit uniaxial, or even biaxial, negative thermal expansion, which accommodates extremely large positive expansion. This preserves the integrity of the crystal lattice and the crystal as a whole during heating/cooling (during which stress is accumulating in the lattice). More importantly, this probably allows absorption of large elastic energy, which is released at some point in the form of mechanical motion of the crystals. The thermosalient mechanism in methscopolamine bromide is a continuous process unlike other thermosalient materials, which are characterized by sharp phase transition during which thermosalient effect abruptly takes places. This is evidenced by the quite broad temperature interval during which the crystals were jumping (more than 10 K), by the lower intensity of jumps compared to, for example, oxitropium bromide, and by the retained integrity of the crystals.

Supplementary Materials: The following are available online at http://www.mdpi.com/2073-4352/8/7/301/s1, Figure S1: Differential scanning calorimetry of scopolamine bromide; Table S1: Summary of Rietveld structure refinement for methscopolamine at *T* = 308 K and *T* = 458 K; Table S2: Atomic coordinates and isotropic displacement parameters of methscopolamine at 308 K. Hydrogen atoms were not refined; Table S3: Atomic coordinates and isotropic displacement parameters of methscopolamine at 458 K. Hydrogen atoms were not refined. Video S1: heating of methscopolamine bromide (heating rate 10 K/min). Video S2: heating of methscopolamine bromide (heating rate 20 K/min). Video S3: cooling of methscopolamine bromide (cooling rate 10 K/min). Video S4: cooling of methscopolamine bromide (cooling rate 20 K/min).

Author Contributions: The study was designed by Ž.S.; Manuscript was written by J.P and Ž.S.; Experimental work was conducted by T.K.; Data were interpreted by Ž.S., J.P., J.F., M.Z., S.C.T., and T.K.

Funding: This work was supported by the Croatian Science Foundation project IP-2014-09-7506.

Conflicts of Interest: The authors declare no conflict of interest.

References

1. Nath, N.K.; Panda, M.K.; Sahoo, S.C.; Naumov, P. Thermally induced and photoinduced mechanical effects in molecular single crystals—A revival. *Cryst. Eng. Comm.* **2014**, *16*, 1850. [CrossRef]
2. Karothu, D.P.; Weston, J.; Desta, I.T.; Naumov, P. Shape-memory and self-healing effects in mechanosalient molecular crystals. *J. Am. Chem. Soc.* **2016**, *138*, 13298–13306. [CrossRef] [PubMed]
3. Sahoo, S.C.; Sinha, S.B.; Kiran, M.S.R.N.; Ramamurty, U.; Dericioglu, A.F.; Reddy, C.M.; Naumov, P. Kinematic and mechanical profile of the self-actuation of thermosalient crystal twins of 1,2,4,5-tetrabromobenzene: A molecular crystalline analogue of a bimetallic strip. *J. Am. Chem. Soc.* **2013**, *135*, 13843–13850. [CrossRef] [PubMed]
4. Commins, P.; Desta, I.T.; Karothu, D.P.; Panda, M.K.; Naumov, P. Crystals on the move: Mechanical effects in dynamic solids. *Chem. Commun.* **2016**, *52*, 13941–13954. [CrossRef] [PubMed]
5. Panda, M.K.; Runčevski, T.; Husain, A.; Dinnebier, R.E.; Naumov, P. Perpetually self-propelling chiral single crystals. *J. Am. Chem. Soc.* **2015**, *137*, 1895–1902. [CrossRef] [PubMed]
6. Panda, M.K.; Runčevski, T.; Chandra Sahoo, S.; Belik, A.A.; Nath, N.K.; Dinnebier, R.E.; Naumov, P. Colossal positive and negative thermal expansion and thermosalient effect in a pentamorphic organometallic martensite. *Nat. Commun.* **2014**, *5*, 4811. [CrossRef] [PubMed]
7. Panda, M.K.; Centore, R.; Causà, M.; Tuzi, A.; Borbone, F.; Naumov, P. Strong and anomalous thermal expansion precedes the thermosalient effect in dynamic molecular crystals. *Sci. Rep.* **2016**, *6*, 29610. [CrossRef] [PubMed]
8. Tarantino, S.C.; Giannini, M.; Carpenter, M.A.; Zema, M. Cooperative Jahn-Teller effect and the role of strain in the tetragonal-to-cubic phase transition in $Mg_xCu_{1-x}Cr_2O_4$. *IUCrJ* **2016**, *3*, 354–366. [CrossRef] [PubMed]
9. Shibuya, Y.; Itoh, Y.; Aida, T. Jumping crystals of pyrene tweezers: Crystal-to-Crystal transition involving π/π-to-CH/π assembly mode switching. *Chem. Asian J.* **2017**, *12*, 811–815. [CrossRef] [PubMed]
10. Takeda, T.; Akutagawa, T. Anisotropic dissociation of π-π stacking and flipping-motion-induced crystal jumping in alkylacridones and their dicyanomethylene derivatives. *Chem. Eur. J.* **2016**, *22*, 7763–7770. [CrossRef] [PubMed]
11. Ohtani, S.; Gon, M.; Tanaka, K.; Chujo, Y. A flexible, fused, azomethine–boron complex: Thermochromic luminescence and thermosalient behavior in structural transitions between crystalline polymorphs. *Chem. Eur. J.* **2017**, *23*, 11827–11833. [CrossRef] [PubMed]
12. Lončarić, I.; Popović, J.; Despoja, V.; Burazer, S.; Grgičević, I.; Popović, D.; Skoko, Ž. Reversible thermosalient effect of N'-2-propylidene-4- hydroxybenzohydrazide accompanied by an immense negative compressibility: Structural and theoretical arguments aiming toward the elucidation of jumping phenomenon. *Cryst. Growth Des.* **2017**, *17*, 4445–4453. [CrossRef]
13. Khalil, A.; Ahmed, E.; Naumov, P. Metal-coated thermosalient crystals as electrical fuses. *Chem. Commun.* **2017**, *60*, 1–4. [CrossRef] [PubMed]
14. Skoko, Ž.; Zamir, S.; Naumov, P.; Bernstein, J. The thermosalient phenomenon. "Jumping crystals" and crystal chemistry of the anticholinergic agent oxitropium bromide. *J. Am. Chem. Soc.* **2010**, *132*, 14191–14202. [CrossRef] [PubMed]
15. Glaser, R.; Shiftan, D.; Drouin, M. The solid-state structures of (–) -scopolamine free base, (–) -scopolamine methobromide, the pseudopolymorphic forms of (–) -scopolamine hydrochloride anhydrate and 1.66hydrate. *Can. J. Chem.* **2000**, *78*, 212–223. [CrossRef]
16. Cliffe, M.J.; Goodwin, A.L. PASCal: A principal axis strain calculator for thermal expansion and compressibility determination. *J. Appl. Crystallogr.* **2012**, *45*, 1321–1329. [CrossRef]
17. Zakharov, B.A.; Michalchuk, A.A.L.; Morrison, C.A.; Boldyreva, E.V. Anisotropic lattice softening near the structural phase transition in the thermosalient crystal 1,2,4,5-tetrabromobenzene. *Phys. Chem. Chem. Phys.* **2018**, *20*, 8523–8532. [CrossRef] [PubMed]

18. Birkedal, H.; Schwarzenbach, D. Observation of uniaxial negative thermal expansion in an organic crystal. *Angew. Chem. Int. Ed.* **2002**, *41*, 754–756. [CrossRef]

19. Krishnan, R.S.; Srinivasan, R.; Devanarayanan, S. *Thermal Expansion of Crystals*; Pamplin, B.R., Ed.; Pergamon Press: Oxford, UK, 1979.

© 2018 by the authors. Licensee MDPI, Basel, Switzerland. This article is an open access article distributed under the terms and conditions of the Creative Commons Attribution (CC BY) license (http://creativecommons.org/licenses/by/4.0/).

crystals

MDPI

Article

Structural Identification of Binary Tetrahydrofuran + O₂ and 3-Hydroxytetrahydrofuran + O₂ Clathrate Hydrates by Rietveld Analysis with Direct Space Method

Yun-Ho Ahn [1], Byeonggwan Lee [2] and Kyuchul Shin [2,*

[1] School of Chemical and Biomolecular Engineering, Georgia Institute of Technology, Atlanta, GA 30318, USA; yunho09@gmail.com
[2] Department of Applied Chemistry, School of Applied Chemical Engineering, Kyungpook National University, Daegu 41566, Korea; happyboy0222@naver.com
* Correspondence: kyuchul.shin@knu.ac.kr; Tel.: +82-53-950-5587

Received: 24 July 2018; Accepted: 14 August 2018; Published: 18 August 2018

Abstract: The structural determination of clathrate hydrates, nonstoichiometric crystalline host-guest materials, is challenging because of the dynamical disorder and partial cage occupancies of the guest molecules. The application of direct space methods with Rietveld analysis can determine the powder X-ray diffraction (PXRD) patterns of clathrates. Here, we conducted Rietveld analysis with the direct space method for the structural determination of binary tetrahydrofuran (THF) + O₂ and 3-hydroxytetrahydrofuran (3-OH THF) + O₂ clathrate hydrates in order to identify the hydroxyl substituent effect on interactions between the host framework and the cyclic ether guest molecules. The refined PXRD results reveal that the hydroxyl groups are hydrogen-bonded to host hexagonal rings of water molecules in the $5^{12}6^4$ cage, while any evidences of hydrogen bonding between THF guests and the host framework were not observed from PXRD at 100 K. This guest-host hydrogen bonding is thought to induce slightly larger 5^{12} cages in the 3-OH THF hydrate than those in the THF hydrate. Consequently, the disorder dynamics of the secondary guest molecules also can be affected by the hydrogen bonding of larger guest molecules. The structural information of binary clathrate hydrates reported here can improve the understanding of the host-guest interactions occurring in clathrate hydrates and the specialized methodologies for crystal structure determination of clathrate hydrates.

Keywords: clathrate hydrate; powder X-ray diffraction; Rietveld refinement

1. Introduction

Clathrate hydrates are nonstoichiometric crystalline host-guest compounds stabilized by van der Waals interactions between hydrogen-bonded water cages and hydrophobic guest molecules [1,2]. Because they have high capacities for gas storage reaching 170 v/v, clathrate hydrates are potentially applicable in the areas of gas storage, separation, transportation, and carbon sequestration [3–6]. In nature, natural gas clathrate hydrates are abundant in permafrost or subsea sediment regions; hence, clathrate hydrates of natural gases are also considered as potential energy resources [7–11]. For application as gas storage materials or as natural gas sources, the understanding of the physicochemical properties of clathrate hydrates, including thermodynamic stability, guest distributions and occupancies, and formation kinetics, is essential. The structural characterization of such materials regarding host-guest interactions is thus a prerequisite for an improved understanding of the inherent nature of clathrate hydrates. However, structural determination of clathrate hydrates is challenging because of dynamical disorder and partial cage occupancies of the guest molecules. In particular, X-ray diffraction analysis for the structural characterization of powdered clathrate hydrate samples is often difficult because clathrate hydrates contain many hydrogen atoms. The contributions of hydrogen atoms to the diffraction patterns are significant, but the atomic scattering factor of hydrogen is too small to use for the analysis of low-resolution powder diffraction patterns.

Recently, Takeya et al. reported that the application of direct space methods with Rietveld analysis can solve powder X-ray diffraction (PXRD) patterns of these clathrate hydrate materials [12]. They suggested that the position of rigid body guest molecules in the cages of the fixed host framework can be determined by a Monte-Carlo approach minimizing reliability factors of refined patterns and also demonstrated that the dynamical disorder of guest molecules in the cages of structure I (sI; cubic *Pm-3n*), structure II (sII; cubic *Fd-3m*), and structure H (sH; hexagonal *P6/mmm*) clathrate hydrates can be refined by the direct-space technique. To overcome the limitation of the small scattering amplitude of hydrogen atoms, they used virtual chemical species with sums of atomic scattering factors instead of refining the hydrogen positions. Shin et al. demonstrated that the hydrogen atom positions of host water molecules can be refined with some distance constraints between oxygen and hydrogen atoms of water molecules when synchrotron high-resolution powder diffraction data were used [13]. Therefore, structural determination of clathrate hydrates including hydrogen atom positions can be achieved by Rietveld refinement analysis with the direct space method of high-resolution PXRD patterns.

Three widely known crystal structures of clathrate hydrates exist [1,2]. The sI hydrate, whose lattice parameter is ~12 Å, contains six tetrakaidecahedrons ($5^{12}6^2$) and two pentagonal dodecahedrons (5^{12}) cages in the unit cell comprising 46 H_2O molecules. The sII hydrate, whose lattice parameter is ~17.3 Å, contains eight hexakaidecahedrons ($5^{12}6^4$) and sixteen 5^{12} cages in the unit cell comprising 136 H_2O molecules. The sH hydrate, with the lattice parameters *a* ~12.2 Å and *c* ~10.1 Å, contains one icosahedron ($5^{12}6^8$), two irregular dodecahedrons ($4^35^66^3$), and three 5^{12} cages in the unit cell comprising 34 H_2O molecules. Some decades ago, canonical clathrate hydrates were thought to be stabilized by van der Waals interactions only, without any directional guest-host interactions [1,14]. However, recent studies have revealed that hydrogen bonding or halogen bonding between the host and guest molecules occasionally occurs in clathrate hydrate phases [14–20].

Here, we conducted Rietveld analysis with the direct space method for the structural determination of binary tetrahydrofuran (THF) + O_2 and 3-hydroxytetrahydrofuran (3-OH THF) + O_2 clathrate hydrates in order to identify the hydroxyl substituent effect on interactions between the host framework and cyclic ether guest molecules. THF is a widely known hydrate-forming cyclic ether that occupies the $5^{12}6^4$ cages of sII hydrate [1,2,21–23]. On the other hand, 3-OH THF, a hydroxyl group substituted THF, cannot form sII hydrate alone but can form a sII hydrate with secondary gaseous guest molecules such as O_2, N_2, or CH_4 [24]. Because of the inhibition effect of the hydroxyl group, binary (3-OH THF + gaseous guest) hydrates are thermodynamically less stable than binary THF hydrates. In this work, high-resolution PXRD patterns of THF + O_2 and 3-OH THF + O_2 binary hydrates were obtained from a synchrotron beam line and refined by the direct space method and Rietveld method in order to investigate the effect of hydroxyl groups on enclathrated cyclic ether guest molecules.

2. Experimental Section

THF and 3-OH THF were supplied by Sigma-Aldrich Inc. (St. Louis, MO, USA) and used without further purification. O_2 gas of 99.95 mol % purity was purchased from Special Gas (Daejeon, Korea).

A well-mixed solution of THF/H_2O or 3-OH THF/H_2O at the mole ratio of 1:17 was prepared. A high-pressure reactor with an internal volume of 50 mL was loaded with 10 g of each solution and then placed in a refrigerated ethanol circulator (RW-2025G, Jeio Tech Co., Ltd., Daejeon, Korea) and pressurized by O_2 up to 12 MPa at 293 K. The fluids inside the reactor were mechanically stirred throughout hydrate formation. After the system reached the steady state, the reactor was slowly cooled to 253 K and maintained at that temperature for three days. The synthesized hydrate samples were collected and finely ground at liquid nitrogen temperature. The powdered samples were kept in a liquid nitrogen storage dewar until diffraction pattern measurement.

The PXRD patterns were obtained using the supramolecular crystallography beamline (2D) at the Pohang Accelerator Laboratory (PAL) in Korea. An ADSC Quantum 210 CCD detector (210 mm × 210 mm) with synchrotron radiation ($\lambda = 0.9000$ Å) was used. A pre-cooled polyimide tube (purchased from Cole-Parmer, Vernon Hills, IL, USA); inner diameter: 0.025 in; outer diameter: 0.269 in) was filled with the powdered hydrate sample and loaded into the diffractometer. The sample-detector distance was 63 mm. Two-dimensional patterns of 4096 pixels by 4096 pixels were recorded with an exposure time of 5 s at 100 K and then were converted into one-dimensional diffraction patterns of 2θ range from 0 to 66.9932°. As the intensities near the starting and endpoints of the patterns diverged, the 2θ regions from 0 to 4 and from 66 to endpoint were excluded. The obtained patterns were refined by Rietveld analysis with the direct space method [12]. The guest molecules of THF, 3-OH THF, and O_2 were assumed as rigid bodies, and their positions were determined by the direct space method using the program FOX [25,26]. With these initial guest coordinates, the patterns were refined by the Rietveld method with the FULLPROF program [27]. During the refinements, soft distance constraints for the host water molecules (O-H covalent bond length: 0.98 Å; and O·H hydrogen bond length: 1.74 Å) were applied. The isotropic atomic displacement parameters for hydrogen atoms can be experientially constrained to be some factor times the values for the atoms to which the hydrogens are bonded [28]. In this work, the isotropic temperature factor (B value, defined as $B = 8\pi^2 \langle u^2 \rangle$ where $\langle u^2 \rangle$ is the mean square isotropic displacement) of a hydrogen atom of H_2O, THF, or 3-OH THF was assumed to be 1.5 times the B value factor of the atom to which the hydrogen was bonded [29]. The B values for the carbon and oxygen atoms of THF or 3-OH THF were defined as identical.

3. Results and Discussion

The atomic coordinates, isotropic temperature factors, and site occupancies of THF + O_2 and 3-OH THF + O_2 hydrates, which were determined by Rietveld analysis with the direct space method, are presented in Tables 1 and 2.

Table 1. Atomic coordinates and isotropic temperature factors for binary tetrahydrofuran (THF) + O_2 hydrate at 100 K. H_{ea}: hydrogen covalently connected with O_e and hydrogen bonded with O_a (O_e-H_{ea}·O_a). H_{gg}(p): hydrogen in pentagonal ring; and H_{gg}(h): in hexagonal ring. Site: multiplicity and Wyckoff letter.

Atom	x	y	z	B (Å^2)	g	Site
O_a	0.125	0.125	0.125	1.84 (10)	1	8 *a*
O_e	0.2166 (1)	0.2166	0.2166	1.56 (6)	1	32 *e*
O_g	0.1822 (1)	0.1822	0.3706 (1)	1.80 (3)	1	96 *g*
H_{ea}	0.1842 (2)	0.1842	0.1842	2.34	0.5	32 *e*
H_{ae}	0.1581 (2)	0.1581	0.1581	2.76	0.5	32 *e*
H_{eg}	0.2106 (10)	0.2106	0.2731 (6)	2.34	0.5	96 *g*
H_{ge}	0.1827 (10)	0.1827	0.3110 (5)	2.70	0.5	96 *g*
H_{gg}(p)	0.1414 (3)	0.1414	0.3640 (19)	2.70	0.5	96 *g*
H_{gg}(h)	0.2373 (7)	0.1807 (8)	0.3912 (11)	2.70	0.5	192 *i*
C_L1	0.9314	0.9142	0.3679	3.43 (29)	0.0415 (2)	192 *i*
O_L2	0.9143	0.8484	0.3197	3.43	0.0415	192 *i*
C_L3	0.8364	0.8600	0.2924	3.43	0.0415	192 *i*
C_L4	0.7909	0.8937	0.3624	3.43	0.0415	192 *i*
C_L5	0.8557	0.9311	0.4138	3.43	0.0415	192 *i*
H_L6	0.9811	0.8990	0.4045	5.14	0.0415	192 *i*
H_L7	0.9474	0.9644	0.3312	5.14	0.0415	192 *i*
H_L8	0.8365	0.9012	0.2433	5.14	0.0415	192 *i*
H_L9	0.8146	0.8039	0.2718	5.14	0.0415	192 *i*
H_L10	0.7471	0.9359	0.3438	5.14	0.0415	192 *i*
H_L11	0.7609	0.8473	0.3941	5.14	0.0415	192 *i*
H_L12	0.8576	0.9038	0.4712	5.14	0.0415	192 *i*
H_L13	0.8469	0.9936	0.4223	5.14	0.0415	192 *i*
O_S1	0.2317	0.2285	0.9789	5.31 (19)	0.0808 (4)	192 *i*
O_S2	0.2775	0.2695	1.0131	5.31	0.0808	192 *i*

Table 2. Atomic coordinates and isotropic temperature factors for 3-hydroxytetrahydrofuran (3-OH THF) + O_2 hydrate at 100 K.

Atom	x	y	z	B (Å^2)	g	Site
O_a	0.125	0.125	0.125	1.42 (9)	1	8 a
O_e	0.2166 (1)	0.2166	0.2166	1.55 (6)	1	32 e
O_g	0.1824 (1)	0.1824	0.3708 (1)	1.90 (3)	1	96 g
H_{ea}	0.1842 (2)	0.1842	0.1842	2.32	0.5	32 e
H_{ae}	0.1578 (2)	0.1578	0.1578	2.13	0.5	32 e
H_{eg}	0.2156 (9)	0.2156	0.2752 (5)	2.32	0.5	96 g
H_{ge}	0.1857 (11)	0.1857	0.3127 (6)	2.85	0.5	96 g
H_{gg}(p)	0.1416 (3)	0.1416	0.3835 (15)	2.85	0.5	96 g
H_{gg}(h)	0.2372 (7)	0.1782 (9)	0.3878 (12)	2.85	0.5	192 i
C_L1	0.6740	0.1037	0.1417	1.89 (34)	0.0387 (2)	192 i
O_L2	0.6493	0.0585	0.0732	1.89	0.0387	192 i
C_L3	0.5662	0.0766	0.0571	1.89	0.0387	192 i
C_L4	0.5426	0.1418	0.1146	1.89	0.0387	192 i
C_L5	0.5986	0.1280	0.1841	1.89	0.0387	192 i
H_L6	0.7121	0.0671	0.1774	2.83	0.0387	192 i
H_L7	0.7047	0.1574	0.1252	2.83	0.0387	192 i
H_L8	0.5601	0.0947	−0.0039	2.83	0.0387	192 i
H_L9	0.5315	0.0237	0.0668	2.83	0.0387	192 i
H_L10	0.5564	0.1996	0.0909	2.83	0.0387	192 i
H_L11	0.4809	0.1392	0.1307	2.83	0.0387	192 i
H_L12	0.5771	0.0795	0.2203	2.83	0.0387	192 i
O_L13	0.6164	0.1972	0.2297	1.89	0.0387	192 i
H_L14	0.5686	0.2118	0.2593	2.83	0.0387	192 i
O_S1	0.2829	0.2292	0.9757	4.27 (25)	0.0785 (5)	192 i
O_S2	0.2440	0.2626	1.0236	4.27	0.0785	192 i

THF is a widely known sII hydrate former. As expected, the pattern of THF + O_2 hydrate sample in Figure 1a shows the cubic *Fd-3m* structure with a lattice parameter of *a* = 17.1143 (5) Å. The calculated density is 1.166 g/cm^3. As reported previously [24], the pattern of the 3-OH THF + O_2 hydrate sample also shows the cubic *Fd-3m* structure with a lattice parameter of *a* = 17.1268(5) Å, with a tiny amount of hexagonal ice impurities (Figure 1b). The calculated density of the hydrate phase is 1.186 g/cm^3. The refined cage occupancy values for THF and 3-OH THF in the $5^{12}6^4$ cages are 1.00 (1) and 0.93 (1), respectively (Tables 1 and 2). The slightly smaller value of the latter system may arise from the nature of the 3-OH THF molecule, which cannot form the sII hydrate on its own. Although 3-OH THF is a larger guest molecule than THF (Figure 2), the estimated average radii (average distances between the cage centers and each oxygen atom) of the $5^{12}6^4$ cages are almost equal at 4.629 Å and 4.630 Å for the THF and 3-OH THF hydrates, respectively. Here, we assumed that the $5^{12}6^4$ cages are occupied by only large hydrocarbon molecules in both the THF and 3-OH THF hydrates in this work. Although a possibility of O_2 occupancy in the $5^{12}6^4$ cages of the 3-OH THF hydrate exists, the number of small guest molecules occupying the large cages is usually ignorable when a stoichiometric amount of large guest molecules for sII hydrate is used.

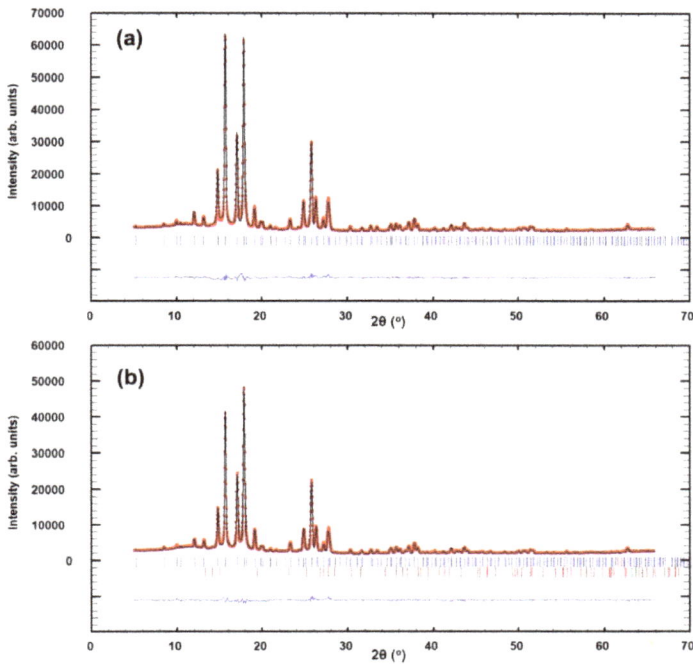

Figure 1. (**a**) Rietveld refinement of the THF + O_2 hydrate pattern. Space group: *Fd-3m*; Lattice parameter: a = 17.1143 (5) Å; Reliability factors: χ^2 = 5.68; and R_{wp} = 8.55% (background subtracted); (**b**) Rietveld refinement of the 3-OH THF + O_2 hydrate pattern (tick marks: first row for sII hydrate, second row for ice I_h). Space group: *Fd-3m*; Lattice parameter: a = 17.1268 (5) Å; Reliability factors: χ^2 = 3.65; and R_{wp} = 8.38% (background subtracted).

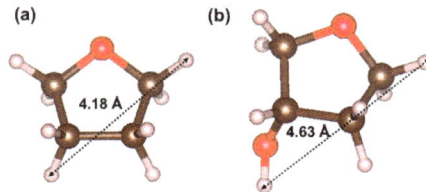

Figure 2. Molecular shapes and the longest end-to-end distances of (**a**) THF and (**b**) 3-OH THF.

The hydroxyl group capable of hydrogen bonding is usually allowed to approach the host water molecules more closely than other hydrophobic groups in the guest molecules [15,20,30,31]; this may explain the 3-OH THF formation of the sII hydrate, while 2-methyl THF, which is similar in size, forms the sH hydrate with the assistance of CH_4 or Xe gases [32].

Figures 3 and 4 show the crystal structures and guest positions of the THF + O_2 and 3-OH THF + O_2 hydrates as obtained by Rietveld refinement.

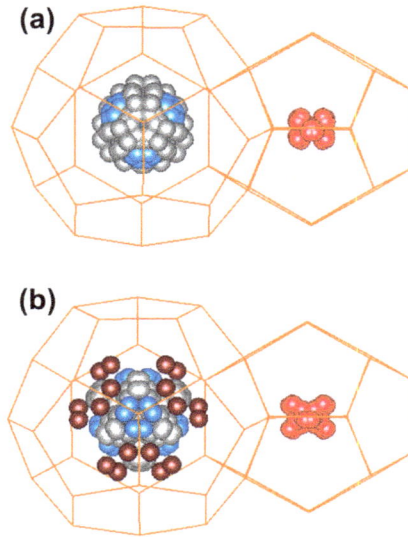

Figure 3. Guest distributions with full symmetry in $5^{12}6^4$ and 5^{12} cages. (**a**) THF + O_2 and (**b**) 3-OH THF + O_2 hydrates. (Red: oxygen of O_2 molecule; Gray: carbon; blue: oxygen in cyclic rings; brown: oxygen of hydroxyl group. Hydrogen atoms are omitted.).

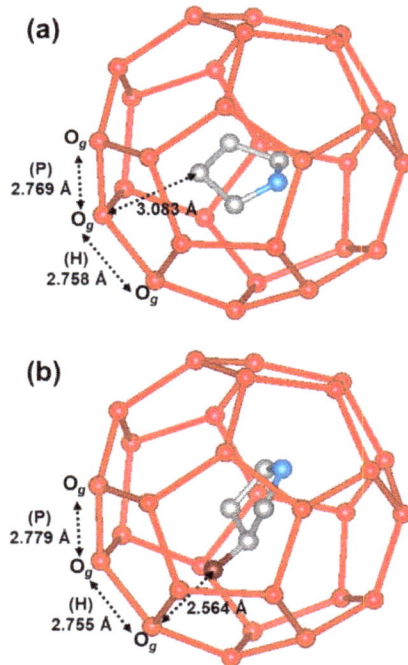

Figure 4. Guest molecules in the $5^{12}6^4$ cages of (**a**) THF + O_2 and (**b**) 3-OH THF + O_2 hydrates.

As shown in Figure 3a, THF molecules are spherically distributed in the $5^{12}6^4$ cages. The shortest distance between a host oxygen atom and guest carbon or oxygen atom is calculated as 3.083 Å (Figure 4a). On the other hand, Figure 3b shows that the hydroxyl groups of 3-OH THF are oriented toward the hexagonal faces of the $5^{12}6^4$ cages. The shortest host-guest distance between oxygen atoms is calculated as 2.564 Å (Figure 4b); the PXRD analysis thus reveals that the hydroxyl functional group of 3-OH THF is hydrogen-bonded to the host water molecules. As H_2O molecules in the sII hydrates are tetrahedrally connected to each other, the hexagonal rings in the $5^{12}6^4$ cages are relatively weakly hydrogen-bonded in the host framework (the O-O-O angles are ~120°, whereas those in pentagonal rings are ~108°). Therefore, the guest-host hydrogen bonding in the $5^{12}6^4$ cage often occurs at the hexagonal rings, as shown in Figures 3b and 4b [18–20].

The distances between host oxygen atoms can be affected by the guest-host hydrogen bonding. In the THF + O_2 hydrate, the O_g-O_g distances in the hexagonal and pentagonal faces are calculated as 2.758 Å and 2.769 Å, respectively (Figure 4). On the other hand, those distances for the 3-OH THF + O_2 hydrate are 2.755 Å and 2.779 Å in the hexagonal and pentagonal faces, respectively. The slightly larger difference in the O_g-O_g distance between the hexagonal and pentagonal faces of the 3-OH THF + O_2 hydrate suggests a slightly distorted framework in the 3-OH THF + O_2 hydrate, caused by the guest-host hydrogen bonding. The host O-O distances are listed in Table 3.

Table 3. Distances of host O-O atoms in the THF + O_2 and 3-OH THF + O_2 hydrates.

Hydrates	O_a-O_e (Å)	O_e-O_g (Å)	O_g-O_g (p) [1] (Å)	O_g-O_g (h) [2] (Å)
THF + O_2	2.714 (2)	2.765 (2)	2.769 (2)	2.758 (3)
3-OH THF + O_2	2.716 (2)	2.768 (3)	2.779 (2)	2.755 (3)

[1] in the pentagonal ring; [2] in the hexagonal ring.

The average radii of the small 5^{12} cages for both hydrates are also estimated; the values are 3.858 Å for the THF hydrate and 3.862 Å for the 3-OH THF hydrate. The 5^{12} cages in the latter are slightly larger than those in the former. This can be a result of framework distortion induced by the significant guest-host hydrogen bonding occurring in the large $5^{12}6^4$ cage of 3-OH THF hydrate. The off-centered distances of the O_2 guest molecules from the centers of the 5^{12} cages are calculated at 0.11 Å and 0.24 Å for the THF and 3-OH THF hydrates, respectively. The estimated 5^{12} cage radii and the off-centered distances support that the position of the O_2 guest molecule in the 3-OH THF hydrate fluctuates more as shown in Figure 3, representing the disorder dynamics of the guest molecules. However, as the mean square isotropic displacement ($\langle u^2 \rangle$) of guest O atoms in THF hydrate, meaning thermal vibration amplitude of O atom, is slightly larger (0.067(2) Å^2 for the THF hydrate and 0.054 (3) Å^2 for the 3-OH THF hydrate) than the value in the 3-OH THF hydrate, it is difficult to conclude that the O_2 guest molecules captured in the 3-OH THF hydrate occupy more space than those in the THF hydrate.

4. Conclusions

In this work, the PXRD patterns of the two binary clathrate hydrates of THF + O_2 and 3-OH THF + O_2 were analyzed by Rietveld refinement with the direct space method. The hydroxyl group of 3-OH THF was hydrogen-bonded to the host water molecules in the hexagonal rings of the $5^{12}6^4$ cages. This guest-host hydrogen bonding slightly distorted the framework and is thought to induce larger 5^{12} cage in the 3-OH THF hydrate. Consequentially, the disorder dynamics of the secondary small guest molecules also can be affected by hydrogen bonding between large guest molecules and the host framework. The findings presented in this work can provide a better understanding of host-guest interactions occurring in clathrate hydrates and the specialized methodologies for the crystal structure determination of clathrate hydrates.

Author Contributions: Conceptualization, K.S. and Y.-H.A.; Methodology, Y.-H.A. and K.S.; Software, K.S.; Validation, K.S., Y.-H.A. and B.L.; Formal Analysis, K.S. and B.L.; Investigation, Y.-H.A.; Resources, K.S.; Data Curation, K.S., Y.-H.A. and B.L.; Writing-Original Draft Preparation, K.S.; Writing-Review & Editing, K.S., Y.-H.A. and B.L.; Visualization, K.S.; Supervision, K.S.; Project Administration, K.S.; Funding Acquisition, K.S.

Funding: This work was supported by the National Research Foundation of Korea (NRF) grant (NRF-2018R1D1A1B07040575) funded by the Ministry of Education (MOE) and the Korea Institute of Energy Technology Evaluation and Planning (KETEP) grant (No. 20141510300310) funded by the Ministry of Trade, Industry & Energy (MOTIE).

Acknowledgments: PXRD experiments were performed at the beamline 2D of the Pohang Accelerator Laboratory (PAL).

Conflicts of Interest: The authors declare no conflicts of interest.

References

1. Jeffrey, G.A. *Inclusion Compounds*; Atwood, J.L., Davies, J.E.D., MacNicol, D.D., Eds.; Academic Press: London, UK, 1984; Volume 1, pp. 135–190.
2. Sloan, E.D.; Koh, C.A. *Clathrate Hydrates of Natural Gases*, 3rd ed.; CRC Press, Taylor & Francis Group: Boca Raton, FL, USA, 2008.
3. Wang, W.; Bray, C.L.; Adams, D.J.; Cooper, A.I. Methane Storage in Dry Water Gas Hydrates. *J. Am. Chem. Soc.* **2008**, *130*, 11608–11609. [CrossRef] [PubMed]
4. Kang, S.P.; Lee, H. Recovery of CO_2 from Flue Gas Using Gas Hydrate: Thermodynamic Verification through Phase Equilibrium Measurements. *Environ. Sci. Technol.* **2000**, *34*, 4397–4400. [CrossRef]
5. Stern, L.A.; Circone, S.; Kirby, S.H.; Durham, W.B. Temperature, Pressure, and Compositional Effects on Anomalous or "Self" Preservation of Gas Hydrates. *Can. J. Phys.* **2003**, *81*, 271–283. [CrossRef]
6. Babu, P.; Kumar, R.; Linga, P. Pre-Combustion Capture of Carbon Dioxide in a Fixed Bed Reactor Using the Clathrate Hydrate Process. *Energy* **2013**, *50*, 364–373. [CrossRef]
7. Koh, C.A.; Sum, A.K.; Sloan, E.D. State of the Art: Natural Gas Hydrates as a Natural Resource. *J. Nat. Gas Sci. Eng.* **2012**, *8*, 132–138. [CrossRef]
8. Ohgaki, K.; Takano, K.; Sangawa, H.; Matsubara, T.; Nakano, S. Methane Exploitation by Carbon Dioxide from Gas Hydrates Phase Equilibria for CO_2-CH_4 Mixed Hydrate System. *J. Chem. Eng. Jpn.* **1996**, *29*, 478–483. [CrossRef]
9. Lee, H.; Seo, Y.; Seo, Y.T.; Moudrakovski, I.L.; Ripmeester, J.A. Recovering Methane from Solid Methane Hydrate with Carbon Dioxide. *Angew. Chem. Int. Ed.* **2003**, *42*, 5048–5051. [CrossRef] [PubMed]
10. Bai, D.; Zhang, X.; Chen, G.; Wang, W. Replacement Mechanism of Methane Hydrate with Carbon Dioxide from Microsecond Molecular Dynamic Simulations. *Energy Environ. Sci.* **2012**, *5*, 7033–7041. [CrossRef]
11. Chong, Z.R.; Yang, S.H.B.; Babu, P.; Linga, P.; Li, X.-S. Review of Natural Gas Hydrates as an Energy Resource: Prospects and Chanllenges. *Appl. Energy* **2016**, *162*, 1633–1652. [CrossRef]
12. Takeya, S.; Udachin, K.A.; Moudrakovski, I.L.; Susilo, R.; Ripmeester, J.A. Direct Space Methods for Powder X-ray Diffraction for Guest-Host Materials: Applications to Cage Occupancies and Guest Distributions in Clathrate Hydrates. *J. Am. Chem. Soc.* **2010**, *132*, 524–531. [CrossRef] [PubMed]
13. Shin, K.; Cha, M.; Lee, W.; Seo, Y.; Lee, H. Abnormal Proton positioning of Water Framework in the Presence of Paramagnetic Guest within Ion-Doped Clathrate Hydrate Host. *J. Phys. Chem. C* **2014**, *118*, 15193–15199. [CrossRef]
14. Susilo, R.; Alavi, S.; Moudrakovski, I.L.; Englezos, P.; Ripmeester, J.A. Guest-Host Hydrogen Bonding in Structure H Clathrate Hydrates. *ChemPhysChem* **2009**, *10*, 824–829. [CrossRef] [PubMed]
15. Alavi, S.; Udachin, K.; Ripmeester, J.A. Effect of Guest-Host Hydrogen Bonding on the Structures and Properties of Clathrate Hydrates. *Chem. Eur. J.* **2010**, *16*, 1017–1025. [CrossRef] [PubMed]
16. Udachin, K.A.; Alavi, S.; Ripmeester, J.A. Water-Halogen Interactions in Chlorine and Bromine Clathrate Hydrates: An Example of Multidirectional Halogen Bonding. *J. Phys. Chem. C* **2013**, *117*, 14176–14182. [CrossRef]

17. Alavi, S.; Takeya, S.; Ohmura, R.; Woo, T.K.; Ripmeester, J.A. Hydrogen-bonding alcohol-water interactions in binary ethanol, 1-propanol, and 2-propanol + methane structure II clathrate hydrates. *J. Chem. Phys.* **2010**, *133*, 074505. [CrossRef] [PubMed]

18. Udachin, K.; Alavi, S.; Ripmeester, J.A. Communication: Single Crystal X-ray Diffraction Observation of Hydrogen Bonding between 1-Propanol and Water in a Structure Ii Clathrate Hydrate. *J. Chem. Phys.* **2011**, *134*, 121104. [CrossRef] [PubMed]

19. Shin, K.; Kumar, R.; Udachin, K.A.; Alavi, S.; Ripmeester, J.A. Ammonia Clathrate Hydrates as New Solid Phases for Titan, Enceladus, and other Planetary Systems. *Proc. Natl. Acad. Sci. USA* **2012**, *109*, 14785–14790. [CrossRef] [PubMed]

20. Shin, K.; Udachin, K.A.; Moudrakovski, I.L.; Leek, D.M.; Alavi, S.; Ratcliffe, C.I.; Ripmeester, J.A. Methanol Incorporation in Clathrate Hydrates and the Implications for Oil and Gas Pipeline Flow Assurance and Icy Planetary Bodies. *Proc. Natl. Acad. Sci. USA* **2013**, *110*, 8437–8442. [CrossRef] [PubMed]

21. Jones, C.Y.; Marshall, S.L.; Chakoumakos, B.C.; Rawn, C.J.; Ishii, Y. Structure and Thermal Expansivity of Tetrahydrofuran Deuterate Determined by Neutron Powder Diffraction. *J. Phys. Chem. B* **2003**, *107*, 6026–6031. [CrossRef]

22. Lee, H.; Lee, J.-W.; Kim, D.Y.; Park, J.; Seo, Y.-T.; Zeng, H.; Moudrakovski, I.L.; Ratcliffe, C.I.; Ripmeester, J.A. Tuning Clathrate Hydrates for Hydrogen Storage. *Nature* **2005**, *434*, 743–746. [CrossRef] [PubMed]

23. Florusse, L.J.; Peters, C.J.; Schoonman, J.; Hester, K.C.; Koh, C.A.; Dec, S.F.; Marsh, K.N.; Sloan, E.D. Stable Low-Pressure Hydrogen Clusters Stored in a Binary Clathrate Hydrate. *Science* **2004**, *306*, 469–471. [CrossRef] [PubMed]

24. Ahn, Y.-H.; Kang, H.; Koh, D.-Y.; Park, Y.; Lee, H. Gas hydrate Inhibition by 3-Hydroxytetrahydrofuran: Spectroscopic Identifications and Hydrate Phase Equilibria. *Fluid Phase Equilib.* **2016**, *413*, 65–70. [CrossRef]

25. Favre-Nicolin, F.; Cerny, R. 'Free objects for crystallography': A Modular Approach to Ab Initio Structure Determination from Powder Diffraction. *J. Appl. Crystallogr.* **2002**, *35*, 734–743. [CrossRef]

26. Cerny, R.; Favre-Nicolin, F. Direct Space Methods of Structure Determination from Powder Diffraction: Principles, Guidelines and Perspectives. *Z. Kristallogr.* **2007**, *222*, 105–113.

27. Rodriguez-Carvajal, J. Recent Advances in Magnetic Structure Determination by Neutron Powder Diffraction. *Phys. B* **1993**, *192*, 55–69. [CrossRef]

28. Massa, W. *Crystal Structure Determination*, 2nd ed.; Springer: New York, NY, USA, 2004.

29. Kirchner, M.T.; Boese, R.; Billups, W.E.; Norman, L.R. Gas Hydrate Single-Crystal Structure Analyses. *J. Am. Chem. Soc.* **2004**, *126*, 9407–9412. [CrossRef] [PubMed]

30. Alavi, S.; Susilo, R.; Ripmeester, J.A. Linking Microscopic Guest Properties to Macroscopic Observables in Clathrate Hydrates: Guest-Host Hydrogen Bonding. *J. Chem. Phys.* **2009**, *130*, 174501. [CrossRef] [PubMed]

31. Alavi, S.; Shin, K.; Ripmeester, J.A. Molecular Dynamics Simulations of Hydrogen Bonding in Clathrate Hydrates with Ammonia and Methanol Guest Molecules. *J. Chem. Eng. Data* **2015**, *60*, 389–397. [CrossRef]

32. Lee, J.-W.; Lu, H.; Moudrakovski, I.L.; Ratcliffe, C.I.; Ripmeester, J.A. Thermodynamic and molecular-scale analysis of new systems of water-soluble hydrate formers+ CH_4. *J. Phys. Chem. B* **2010**, *114*, 13393–13398. [CrossRef] [PubMed]

© 2018 by the authors. Licensee MDPI, Basel, Switzerland. This article is an open access article distributed under the terms and conditions of the Creative Commons Attribution (CC BY) license (http://creativecommons.org/licenses/by/4.0/).

crystals

MDPI

Article

Application of Evolutionary Rietveld Method Based XRD Phase Analysis and a Self-Configuring Genetic Algorithm to the Inspection of Electrolyte Composition in Aluminum Electrolysis Baths

Igor Yakimov [1], Aleksandr Zaloga [1,*], Petr Dubinin [1], Oksana Bezrukova [1], Aleksandr Samoilo [1], Sergey Burakov [2], Eugene Semenkin [2], Maria Semenkina [2] and Eugene Andruschenko [1]

[1] Department of Research, Siberian Federal University, 660041 Krasnoyarsk, Russia; i-s-yakimov@yandex.ru (I.Y.); dubinin-2005@yandex.ru (P.D.); opiksina@gmail.com (O.B.); x_lab@rambler.ru (A.S.); andryushchenkoeugene@gmail.com (E.A.)

[2] Department of Systems Analysis and Operations Research, Siberian State University of Science and Technology, 660037 Krasnoyarsk, Russia; burakov_krasu@mail.ru (S.B.); eugenesemenkin@yandex.ru (E.S.); semenkina88@mail.ru (M.S.)

* Correspondence: zaloga@yandex.ru; Tel.: +7-913-559-4935

Received: 30 August 2018; Accepted: 22 October 2018; Published: 24 October 2018

Abstract: The technological inspection of the electrolyte composition in aluminum production is performed using calibration X-ray quantitative phase analysis (QPA). For this purpose, the use of QPA by the Rietveld method, which does not require the creation of multiphase reference samples and is able to take into account the actual structure of the phases in the samples, could be promising. However, its limitations are in its low automation and in the problem of setting the correct initial values of profile and structural parameters. A possible solution to this problem is the application of the genetic algorithm we proposed earlier for finding suitable initial parameter values individually for each sample. However, the genetic algorithm also needs tuning. A self-configuring genetic algorithm that does not require tuning and provides a fully automatic analysis of the electrolyte composition by the Rietveld method was proposed, and successful testing results were presented.

Keywords: x-ray powder diffraction; rietveld method; quantitative XRD phase analysis; aluminum electrolyte; cryolite ratio; genetic algorithms; self-configuration

---

1. Introduction

Aluminum is normally produced by the electrolysis of alumina in molten fluorides at a temperature of around 950 °C. The main component of the molten electrolyte is cryolite (Na_3AlF_6), whilst aluminum fluoride, calcium fluoride, and sometimes magnesium fluoride and potassium fluoride are added to improve the cryolite's technological properties. During the electrolysis, the composition of the electrolyte in the baths continuously changes and shifts from the optimum. The maintenance of an optimal bath composition is a vital element in electrolysis technology. An integral characteristic of the bath composition is the cryolite ratio (CR)—the ratio of molar concentrations of sodium fluoride and aluminum fluoride (1):

$$CR = \frac{C(NaF, \text{ mol. } \%)}{C(AlF_3, \text{ mol. } \%)} = 2 \cdot \frac{C(NaF, \text{ mass. } \%)}{C(AlF_3, \text{ mass. } \%)} \tag{1}$$

The express process control of the electrolyte composition is generally performed by X-ray diffraction quantitative phase analysis (QPA), which uses calibration curves. The cryolite ratio is calculated according to the Equation (1). The concentrations of NaF and AlF_3 are calculated using the results of the QPA of crystallized bath samples. The phase concentrations, in turn, are calculated from the measured intensities of their diffraction peaks. The optimal frequency of measuring the CR is once every two days, the accuracy of the analysis is $\Delta(p = 0.95)\sim 0.04$, and the optimal measurement time per sample is several minutes.

X-ray diffractometers need periodic calibration using electrolyte reference materials with well-established phase composition [1,2]. However, it is difficult to create such reference materials, because, firstly, they must contain all the phases from Table 1. Secondly, the crystal structures of the phases in reference materials must match those in the electrolyte samples. Therefore, QPA by the Rietveld method is more suitable for the process control of bath composition because this method works without using reference materials (as is shown in References [3,4]), yet the low automation of this method in working with such complicated samples limits its applicability. The issue is to set such appropriate initial approximations of both the profile and structural parameters of phases that can be quickly refined automatically by the least squares method (LSM). To address this problem, we suggest applying a genetic algorithm which we have developed to set the initial values of the parameters for each sample automatically [5]. This approach provides high accuracy of measuring the cryolite ratio, but it is not yet fully applicable to the process control. This is because the algorithm must be configured as well. In this paper, we provide a self-configuring genetic algorithm, which works without a preliminary adjustment and performs the fully automated analysis of aluminum electrolyte composition by the Rietveld method.

Table 1. The phase composition of typical industrial bath samples at Russian aluminum smelters.

#	Phase	Chemical Formula	Concentration Range, wt. %	CR Range
1	Cryolite	Na_3AlF_6	0–90	>1.67
2	Chiolite	$Na_5Al_3F_{14}$	0–85	<3.0
3	Sodium fluoride	NaF	0–5	>3.0
4	Ca-cryolite 1	$NaCaAlF_6$	0–15	<3.0
5	Ca-cryolite 2	$Na_2Ca_3Al_2F_{14}$	0–20	<2.95
6	Fluorite	CaF_2	0–9	>2.45
7	Weberite	Na_2MgAlF_7	0–15	<2.85
8	Neiborite	$NaMgF_3$	0–6	>2.5
9	α-, β-, γ-alumina	Al_2O_3	2–5	

References [3,4] also describe an automated analysis of aluminum electrolyte samples using the Rietveld method. However, in these articles, a maximum of 5-phase samples of a calcium-containing electrolyte with an insignificant content (about 0.8%) of the semi-amorphous $NaCaAlF_6$, which difficult to simulate by the Rietveld method, are investigated. As shown below, the proposed approach allows analyses of 8-phase calcium- and magnesium-containing electrolytes (where a feature of Russian aluminum production is adding magnesium fluoride up to 4–5%), with a noticeable $NaCaAlF_6$ content (up to 8% rel.) in an automatic mode for a comparable time. Moreover, in calcium-containing electrolytes, magnesium can also accumulate over time from alumina.

2. Materials and Methods

2.1. The Method of Genetic Algorithm Self-Configuring

Full-profile QPA based on the Rietveld method is widely used for quantitative XRD analysis in laboratories. However, its use for technological inspection in the industry is not yet developed as the Rietveld method is based on a non-linear LSM convergence, which requires a very good approximation of the initial estimations of parameters to be tuned for each sample. In the case of laboratory investigations, the requirements for the initial approximation are not as strict as there exists the possibility for an interactive refinement. However, for a technological inspection in the industry, a high level of automation and the ability to unify the analysis of a large number of samples are strongly required. In this case, the unified initial estimations of a large amount of profile and structure parameters do not fit well to all of them. This results in a divergence in the LSM. One of the possible approaches to tackle this problem is the application of genetic algorithms (GAs) for the choice of an initial approximation of sample parameters, for the evolutionary selection of perspective parameters, and for their automated refinement with Rietveld's LSM.

The application effectiveness of evolutionary algorithms (GAs in particular) depends on the choice of genetic operators: Selection, recombination (crossover), mutation, and substitution. However, the settings for an effective algorithm which ensure that acceptable results are obtained within the shortest possible time can be different for different problems, i.e., they cannot be determined in advance for all cases. Therefore, procedures of dynamic self-adapting and self-configuring of algorithm settings (e.g., References [6,7]) are used here. Self-configuring is an automated choice of effective genetic operators from a given set during an algorithmic run while solving the problem in hand. The configuration of operators is determined stochastically based on the probability of an operator to be used for a generating new solution. These probabilities are calculated according to their success in previous stages. The deployment probability of the most successful operator, the one that gave the best solutions on the previous generation, is increased, whereas the probabilities of other operators are decreased. It makes possible the automated choice of the best configuration of operators for increasing algorithm productivity.

The main stages of a self-configuring genetic algorithm (SGA) for unconditional optimization can be described as follows:

1. Initially, the choice of any particular variant for each kind of operator (selection, crossover, mutation) is equiprobable. More specifically, the probability of choosing a variant of an operator is equal to $p = 1/z$, where z is the number of operator variants. It means that all variants of all operators are used equiprobably before statistics of their effectiveness are collected.

2. On each generation, an effectiveness estimation is performed for each variant of each operator. It is based on the mean fitness of solutions obtained with the use of this variant of this operator: $averagefitness_i = f_i/n_i$, $i = 1, 2, \ldots, z$, where $averagefitness_i$ is the mean fitness of solutions obtained with the i-th variant of the operator; f_i is the fitness sum of all solutions obtained with the i-th variant of the operator; n_i is the number of solutions obtained with the i-th variant of the operator; $i = 1, \ldots, z$, where z is the number of operator variants.

3. For the next generation, the probability of using the most effective variant is increased by $((z-1)\cdot K)/(z\cdot N)$ and probabilities of all other variants are decreased by $K/(z\cdot N)$, where N is the number of established generations of an algorithm run, K is a constant (usually equal to 2 for the considered problems). However, the probability of all variants cannot be lower than a given threshold, whereas the sum of all variant probabilities must be equal to 1. When an operator variant reaches this threshold, it will stop giving out part of its probability, and the best variant will no longer receive it. It is organized in this way because of the possibility that a variant could be unsuccessful on the first stages but could be very useful later on, and this could not happen if its probability decreased to zero.

4. Operators used for the generation of a new solution are chosen stochastically according to obtained probability distributions.

Such self-configuring frees the end user who is not an expert in evolutionary optimization from choosing the settings of the genetic algorithm, whilst the efficiency of solving the problem remains acceptable (with the best choice of the genetic algorithm parameters, the efficiency of solving the problems is somewhat higher, but the selection of the GA parameters requires time and a highly-qualified user).

The effectiveness of an SGA can be improved by using it within the framework of the island evolutionary (cooperative-competing) model, when several populations exist separately from each other, only at times exchanging genetic material. This ensures a more uniform distribution of possible solutions to the problem within the search space. Therefore, in order to solve the problem of quantitative X-ray phase analysis, the following realization of self-configuration of the multi-population parallel genetic algorithm was created. It generates n different populations from the models of the substance being determined, and on each of the n computational nodes of the multi-core personal computers (PC), an individual single-population SGA is run. At the beginning of the process, random individuals are generated, i.e., sets of numbers consisting of the values of the refined parameters of the Rietveld method for the generated models, which are distributed over the search space. At each of the computational nodes, with the help of the recombination and selection operators, evolutionarily occurs the formation of descendants with smaller objective function values. The mutation operators randomly "scatter" them around the search space, sometimes with an increase in the value of the objective function. A proportion of the models with a smaller objective function value are refined using Rietveld's LSM method. Then, as a result of the general selection, a new population of test models is formed, i.e., descendants, on average, with better suitability. A certain number of the best test models from the populations at work nodes are sent to the control computer node of the SGA. All these decisions accumulated on the controlling node are sorted in decreasing order of the value of the objective function. Periodically, some of the best solutions accumulated at a given generation of evolution on the control node are randomly selected and randomly returned to the population at work nodes. Such a moderate migration ensures the spread of successful solutions to populations and improves overall convergence.

Self-configuring is realized for individual GA processes on work units in the way described above. The standard set of genetic operators is given for each process. They are one-point, two-point, and uniform crossover, rank-based and tournament with different size of tournament selection, and low, average, and high selection. Probability redistribution of all operators is performed locally for each work unit irrespective of their effectiveness on other units. This last point could improve the general effectiveness of the algorithm but requires a separate careful study.

2.2. Full-Profile QPA by Parallel Self-Configuring Genetic Algorithms

The essence of the QPA by the Rietveld method is an iterative minimizing of the difference between an experimental powder pattern and the calculated one by the LSM:

$$\Phi(\overline{P}_{k+1}) = \sum_i w_i (Yo(2\Theta_i) - Yc(\overline{P}_k + \Delta\overline{P}_k, 2\Theta_i))^2 \to 0, \tag{2}$$

where Yo, Yc is an experimental and a calculated intensity at a position $2\theta i$, respectively; w_i is a weight coefficient; $\overline{P}_k$ is the vector of profile, microstructural, and structural parameter values at an iteration k; $\Delta\overline{P}_k$ is the parameter increments calculated by the LSM; the initial approximations are set at $k = 0$.

A refinable part of $\overline{P}$ composes parametrical strings that play the role of individuals that are evolutionarily optimized by a GA. The full set of parameters $\overline{P}$, which includes both refinable and fixed parameters, describes a trial model of multiphase sample characteristics. In the case of the evolutionary QPA, a range must be defined within which possible values of refinable parameters fall. The best values found within the range by the GA are then refined by the Rietveld method.

A QPA feature is that it allows the GA to conduct the search for appropriate initial values of parameters within wide ranges. For example, the crystal structures of phases having been found in a sample may be used to set initial values of refinable structural parameters. Crystal structures are normally taken from crystal structure databases. In this case, the atomic coordinates of general crystallographic positions may be chosen as refinable parameters. In addition, the occupation of the positions may be refined for solid solutions. Thus, the range limits the variation of both atomic coordinates and occupation coefficients.

After the completion of the GA process, the final solution is refined using the Rietveld method, and the phase concentrations are calculated using the found parameter values. If the sample does not contain an amorphous phase, the concentrations are calculated according to the following equation:

$$C_\alpha = S_\alpha Z_\alpha M_\alpha V_\alpha / \sum_{j=1}^{N} S_j Z_j M_j V_j, \tag{3}$$

where S_a is the scale factor of a phase a, which is obtained from the calculated powder pattern Yc_a, V_a is the cell volume, Z_a is the number of structural units per cell, M_a is the molecular weight of a phase a, and N is the number of crystalline phases in the sample.

If an amorphous phase is present in the sample, the QPA uses an internal standard [8].

Narrow search ranges pose a problem for the evolutionary full-profile QPA. In such cases, the values of the R-factor vary insufficiently. Therefore, it becomes an unreliable selection criterion. To improve the sensitivity of the criterion, we suggest adding bias between the measured sample's elemental composition and the composition calculated from the phase concentrations.

$$R_{wp} = 100 \cdot \left\{ (1 - w_{Ch}) \cdot \sqrt{\frac{\sum_i w_i \cdot [Yo_i - Yc_i(\overline{P})]^2}{\sum_i w_i \cdot (Yo_i)^2}} + w_{Ch} \cdot \frac{\sum_t w_t \cdot [\sum_\alpha p_{t\alpha} \cdot \frac{S_\alpha Z_\alpha M_\alpha V_\alpha}{\sum_\alpha S_\alpha Z_\alpha M_\alpha V_\alpha} - C_t^{Ch}]^2}{\sum_t w_t \cdot (C_t^{Ch})^2} \right\}, \tag{4}$$

where $C_t{}^{Ch}$ is the concentration of an element t measured by chemical analysis; P_{ta} is the mass fraction of an element t in a phase a; w_{Ch} is the weight contribution of the chemical data in R_{wp} (normally 0.5).

The combination of the suggested variant of the full-profile Rietveld method with the parallel SGA provides an automated QPA. The SGA uses the profile R-factor as a figure of merit to ensure a proper selection of trial models. To perform optimization by the SGA, the special software was written in C++ language. The ObjCryst++ library [9] was used for crystallographic calculations and Rietveld method refinements.

2.3. Objects of Investigations

We applied the SGA to the QPA of the aluminum bath electrolyte. As the model objects, we had chosen 24 branch reference materials used at five Russian aluminum smelters.

The objects had been chosen for the following reasons. Firstly, they were made from real industrial baths taken at different aluminum smelters. Therefore, they were entirely consistent with all the features of real crystallized bath samples, such as composition, impurities, and microstructure. Secondly, the balance between the chemical and phase composition of the reference samples is fully guaranteed by the correspondence among the results obtained by the different analytical methods used for the certification. The mean uncertainty of the certified CR values was 0.008. Thirdly, the quantitative phase composition significantly varies from sample to sample and covers the range of cryolite ratios from 1.9 to 3.

The samples were ground manually in an agate mortar and then pressed in cuvettes from the front side. The powder patterns were obtained with a Shimadzu-7000 powder diffractometer with scintillator detector using $CuK\alpha$ radiation in the range $10° \le 2\Theta \le 90°$; the exposition step was 0.01°. The structural models were taken from the Inorganic crystal structure database (ICSD) [10].

Since the SGA QPA is fully automated, we preset the SGA for the analysis of each sample identically. We also compiled a standard excessive list of contained phases, which is provided in Table 1, for each sample. The nineteen parameters listed in the Table 2 were refined for the chosen phases. In cases when a phase was absent in a sample, its scale factor and the concentration were set to zero. The SGA was run three times for each sample. As the QPA result, we accepted the arithmetic mean of the phase concentrations that were established over three runs. The analysis of each sample lasted about five minutes.

Table 2. Parameters that were refined by the self-configuring genetic algorithm (SGA) for the bath reference materials.

Phase	List of Refinable Parameters
Na_3AlF_6	$S, a, b, c, \beta, U, W, Eta0, Asym1$, atomic coordinates (12 atomic positions), the texture *hkl* [220] by March-Dollase model
$Na_5Al_3F_{14}$	$S, a, c, U, W, Eta0, Asym1$, atomic coordinates (9 atomic positions)
$NaCaAlF_6$	$S, a, b, c, \beta, U, W, Eta0, Asym1$
$Na_2Ca_3Al_2F_{14}$	$S, a, U, W, Eta0, Asym1$
CaF_2	$S, a, U, W, Eta0, Asym1$
Na_2MgAlF_7	$S, a, b, c, U, W, Eta0, Asym1$
$NaMgF_3$	$S, a, b, c, U, W, Eta0, Asym1$
$\alpha\text{-}Al_2O_3$	$S, a, b, U, W, Eta0, Asym1$

where S is the scale factor; a, b, c, β are unit cell parameters; U, W is a peak FWHM by Pseudo-Voigt; $Eta0$ is the peak shape parameter; $Asym1$ is the peak asymmetry parameter.

The research laboratories at aluminum smelters are equipped with combined XRD-XRF analyzers. Normally, the analyzers combine an X-ray diffractometer with a fixed x-ray fluorescence channel that provides quantification of calcium and magnesium. For this reason, we used concentrations of these two elements to calculate the R-factor according to Reference [4].

We preset the following genetic operators for the SGA:

1. Tournament selection among 3, 5, 7, 9 trial models, range selection;
2. Two-point crossover, three-point crossover, uniform crossover;
3. Low-level mutation, average-level mutation, high-level mutation, with three standard deviations each.

The probability of operators varied adaptively for each sample during the SGA performance. In addition, the SGA used a local optimization by Lamarck.

3. Results

Figure 1 shows how the parallel SGA typically converges during the search for the profile and structural parameters. The graph was plotted during the running of the full-profile QPA of a bath reference material. The X-axis shows the number of the generation, whilst the Y axis provides the best corresponding R-factor value that was found among the population at the managing unit. At the zero generation, the SGA randomly generates the initial populations of the trial models.

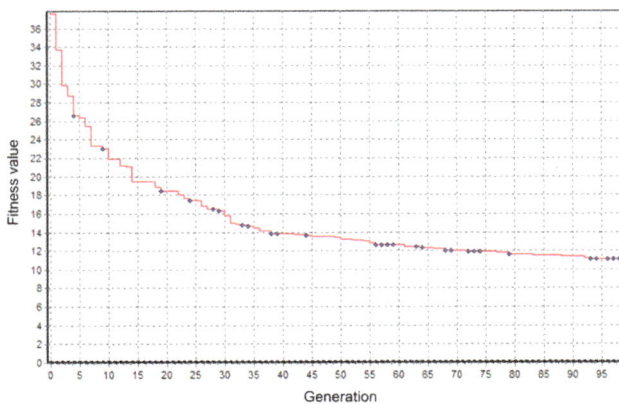

Figure 1. A typical graph of how the SGA converges when analyzing an electrolyte sample.

At the final stage of the QPA, the approximate parameter values, which have been found by the SGA, are exposed to the Rietveld refinement. Figure 2 depicts the experimental powder pattern of the analyzed bath reference sample and the profile that was calculated after the Rietveld refinement. The value of the profile R-factor, which characterizes the difference between the profiles, is 8.6%.

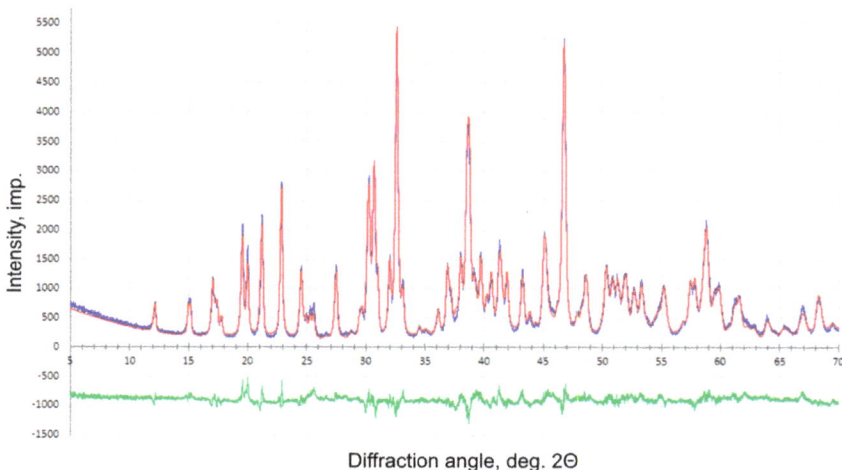

Figure 2. The model powder diffraction pattern calculated by the SGA (in red) and the experimental powder diffraction pattern (in blue) of a bath sample. The green line shows the difference between the profiles.

We propose the correspondence between the certified and calculated values of the cryolite ratio as the quality criterion for results of the evolutionary full-profile QPA. The cryolite ratios were calculated according to the Equation (1). We used the phase concentrations that were computed according to Equation (3) to find the shares of sodium fluorite and aluminum fluorite. Figure 3 shows the correspondence between the certified and calculated values of the cryolite ratio.

Figure 3 also provides the linear regression equation ($y = a + bx$) and the standard deviation, which numerically characterize the correspondence. Therefore, the bias of b from 1 characterizes the systematic error of the results, while the standard deviation describes the random error.

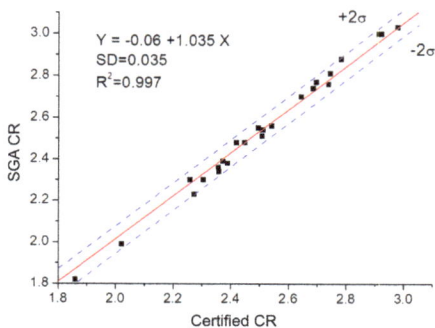

Figure 3. The correspondence between the calculated and certified CR values for the branch reference materials. Certified CR is the certified values; SGA CR is the calculated values; SD is the standard deviation.

Crystals **2018**, *8*, 402

4. Discussion

The calculated values match the certified values with an accuracy of SD = 0.035, and all the results fall within the 95% confidence interval. The linear regression equation is close to a $y = x$ form because the a coefficient is statistically insignificant. However, the b coefficient is 3.5 relative per cent higher than 1. This indicates that the results of the evolutionary full-profile QPA are slightly overestimated.

An analysis shows that this systematic error is caused by the overestimation of the Na_3AlF_6 concentration. This fact proves that the automated Rietveld method that uses SGA data refines the structure of this phase ineffectively. It appears that the structure distorts due to the incomplete transition of the Na_3AlF_6 high-temperature modification to the low-temperature modification. Such an effect is a result of the nonequilibrium crystallization of bath samples, which is caused by a specific sampling procedure being used at aluminum smelters. In addition, the structural distortion inflates the standard deviation of the results.

Overall, the results meet the technological requirements that are set for the accuracy of CR analysis at the smelters. Therefore, we recommend the automated evolutionary method of QPA for the express control of bath composition. However, prior to implementing the method in industry, we must improve its performance by eliminating the causes of the systematic error.

Author Contributions: Conceptualization, I.Y.; Data curation, S.B.; Formal analysis, E.S.; Funding acquisition, I.Y. and A.Z.; Investigation, A.S. and E.A.; Methodology, S.B. and M.S.; Project administration, I.Y.; Resources, O.B. and A.S.; Software, A.Z.; Supervision, P.D. and E.S.; Validation, S.B.; Writing—original draft, I.Y.; Writing—review & editing, A.Z. and P.D.

Funding: The reported study was funded by Russian Foundation for Basic Research, Government of Krasnoyarsk Territory, Krasnoyarsk Regional Fund of Science to the research project No 18-43-243009 Coevolutionary modelling of an atomic-crystal structure of new substances by diffraction data on basis of parallel genetic algorithms and supercomputer computations.

Conflicts of Interest: The authors declare no conflict of interest. The funders had no role in the design of the study; in the collection, analyses, or interpretation of data; in the writing of the manuscript, or in the decision to publish the results.

References

1. *Electrolytic Bath Standards*; Alcan International Ltd.: Montreal, QC, Canada, 2005.
2. Yakimov, I.S.; Dubinin, P.S.; Zaloga, A.N.; Piksina, O.E.; Kirik, S.D. Development of industry standard electrolyte samples of aluminum electrolyzers. *Stand. Samples* **2008**, *4*, 34–42. (In Russian)
3. Feret, F.R. Breakthrough in analysis of electrolytic bath using Rietveld-XRD method. In Proceedings of the TMS2008: 137th Annual Meeting & Exhibition of the Minerals, Metals & Materials Society, New Orleans, LA, USA, 9–13 March 2008; pp. 343–346.
4. Knorr, K.; Kelaar, C. Automated analysis of aluminium bath electrolytes by the Rietveld method. *Miner. Eng.* **2009**, *22*, 434–439. [CrossRef]
5. Zaloga, A.; Akhmedova, A.; Yakimov, I.; Burakov, S.; Semenkin, E.; Dubinin, P.; Piksina, O.; Andryushchenko, E. Genetic Algorithm for Automated X-ray Diffraction Full-Profile Analysis of Electrolyte Composition on Aluminium Smelters. In Proceedings of the Informatics in Control, Automation and Robotics 12th International Conference, Colmar, Alsace, France, 21–23 July 2015; pp. 79–93. [CrossRef]
6. Semenkin, E.; Semenkina, M. The Choice of Spacecrafts' Control Systems Effective Variants with Self-Configuring Genetic Algorithm. In Proceedings of the 9th International Conference on Informatics in Control, Automation and Robotics, ICINCO'2012, Rome, Italy, 28–31 July 2012; Ferrier, J.-L., Bernard, A., Gusikhin, O., Madani, K., Eds.; pp. 84–93.
7. Semenkin, E.S.; Semenkina, M.E. Self-configuring Genetic Algorithm with Modified Uniform Crossover Operator. *Adv. Swarm Intell. Lect. Notes Comput. Sci.* **2012**, *1*, 414–421. [CrossRef]
8. Bish, D.L.; Howard, S.A. Quantitative phase analysis using the Rietveld method. *J. Appl. Cryst.* **1988**, *21*, 86–91. [CrossRef]

9.   Favre-Nicolin, V.; Cerny, R. FOX, 'free objects for crystallography': A modular approach to ab initio structure determination from powder diffraction. *J. Appl. Cryst.* **2002**, *35*, 734–743. [CrossRef]
10. Inorganic Crystal Structure Database. Available online: http://www2.fiz-karlsruhe.de/icsd_publications.html (accessed on 28 August 2018).

© 2018 by the authors. Licensee MDPI, Basel, Switzerland. This article is an open access article distributed under the terms and conditions of the Creative Commons Attribution (CC BY) license (http://creativecommons.org/licenses/by/4.0/).

MDPI

St. Alban-Anlage 66

4052 Basel

Switzerland

Tel. +41 61 683 77 34

Fax +41 61 302 89 18

www.mdpi.com

Crystals Editorial Office

E-mail: crystals@mdpi.com

www.mdpi.com/journal/crystals

www.ingramcontent.com/pod-product-compliance
Lightning Source LLC
Chambersburg PA
CBHW051917210326
41597CB00033B/6168